TALES FROM A
TOP FUEL DRAGSTER

A COLLECTION OF THE GREATEST DRAG RACING STORIES EVER TOLD

BY
SHIRLEY MULDOWNEY

WITH
BILL STEPHENS

FOREWORD BY
DON PRUDHOMME

Copyright © 2005, 2013 by Shirley Muldowney

All Rights Reserved. No part of this book may be reproduced in any manner without the express written consent of the publisher, except in the case of brief excerpts in critical reviews or articles. All inquiries should be addressed to Sports Publishing, 307 West 36th Street, 11th Floor, New York, NY 10018.

Sports Publishing books may be purchased in bulk at special discounts for sales promotion, corporate gifts, fund-raising, or educational purposes. Special editions can also be created to specifications. For details, contact the Special Sales Department, Sports Publishing, 307 West 36th Street, 11th Floor, New York, NY 10018 or sportspubbooks@skyhorsepublishing.com.

Sports Publishing® is a registered trademark of Skyhorse Publishing, Inc.®, a Delaware corporation.

Visit our website at www.sportspubbooks.com

10 9 8 7 6 5 4 3 2 1

Library of Congress Cataloging-in-Publication Data is available on file.
ISBN: 978-1-61321-408-4

Printed in the United States of America

Contents

Foreword vii
Introduction xi

Chapter 1
In the Beginning............................. 2

Chapter 2
On the Strip 14

Chapter 3
The Characters on the Circuit 78

Chapter 4
The Leading Lady 136

Epilogue 168

Appendix
Career Statistics 170

Shirley Muldowney Collection

Shirley Muldowney Collection

Foreword

When you hear the name Shirley Muldowney, there are certain adjectives that instantly pop into your mind to describe the one person who arguably had the biggest impact on the sport of drag racing. Words such as *determined*, *stern*, *passionate*, and *driven* certainly describe her, but I think of Shirley as one of the toughest competitors to ever strap into a Top Fuel car, if not the toughest.

What really stands out to me about Shirley is her perseverance. I've raced against her for a long time and known her even longer. The way she carried on against all the adversity she faced is quite amazing. Just to deal with egotistical males like Don Garlits, Connie Kalitta, Richard Tharp and me was a feat in itself, but to battle back from her injuries and accidents really shows her love and dedication to this sport.

I remember back to the U.S. Nationals in Indianapolis in the 1970s and Shirley was competing in Funny Car and she had a pretty bad fire. I was at the top end at the time and helped the Safety Safari pull her from the burning car. When she took her helmet off and we were sitting there, she had to pull her eyelash off because it was getting into her eye. I remember laughing, and then she laughed too. Shirley might have been out there competing with the boys, but she still had her feminine side.

Following her accident in Montreal in 1984, I visited her in the hospital in Michigan, and to see her come back after the incident—in my opinion—puts her near the top of the list as one of the elite competitors in the history of this sport. That accident solidified our belief that she was truly one of the greats, because many of us, including me, probably would have packed it in after that crash. Why did she come back? I think because she felt she still had something to prove, and that's something I've always admired in her.

Earlier this year, I was in Talladega, Alabama, when Shirley was inducted into the International Motorsports Hall of Fame. It was a great evening. I don't know if she appreciated it as much as we all did. Bob Frey introduced her, and it was a shame all of her fans weren't there to witness it, because he explained her career in a way nobody else could. It was the best induction speech I've ever heard. There wasn't a dry eye in the entire place.

Shirley and I shared many great moments in the sport of drag racing. We struggled together, we thrived together, we rejoiced together, and we shared many personal moments with each other. It was all part of being there at the beginning of the sport.

I think every young lady who comes along in the sport of drag racing and in all of motorsports should tip her hat every time she sees Shirley. She made it possible for them to be successful in racing. Shirley opened the door for females to be accepted and respected in racing. She gave people hope.

Shirley won Summernationals in 1977, making her the first racer to win three consecutive Top Fuel national events. Don Prudhomme is on the left. *Shirley Muldowney Collection*

I have the same respect for her as I do Don Garlits. She was that important to our sport. We all owe her a little something, and hopefully it's respect. The respect for all the great things she has achieved on and off the racetrack is something money can't buy.

—Don Prudhomme

In 2001, the NHRA announced the sports' 50 greatest drivers. Shirley Muldowney was ranked No. 5. Here she is with Don Prudhomme, Don Garlits, Bob Frey, Kenny Bernstein, and John Force. *Shirley Muldowney Collection*

Introduction

Thirty years ago, the sport of NHRA drag racing was a far different animal than it is today. On just about every level, there are vast differences between the way drag racing was conducted in the 1970s and how it goes about its business in this era. Of course, the first thing that comes to mind is the enormous change in the amount of money that was required to compete on the professional tour in the 1970s and the bankroll needed to be competitive—especially in the nitro categories of Top Fuel and Funny Car—in this day and age.

But in a totally different realm, drag racing was far less multicultural, multigenerational, or gender-inclusive as it is now. It was a sport underpinned by the blue-collar work ethic that was defined almost exclusively by men, predominantly white, who were just beginning to raise families while pursuing racing more as a hobby than a career.

It was a world that was familiar to them; a conclave of mostly amateur racers gamboling from place to place, hoping for a trophy or a cash payout, and perhaps, a mention in one of the several hot-rodding magazines favored by the country's motorheads. It was essentially a "guy thing."

Then, Shirley Muldowney changed everything.

Shirley wasn't the first woman to drag race head to head with men. In fact, by the time Shirley arrived on the professional drag racing scene in the mid-1970s, female racers such as Paula Murphy, Judy Lilly, and Shirley Shahan had each made their own individual attempts at breaking the sport's gender barrier. But Shirley not only broke that barrier, she obliterated it. And she did it with a sassy combination of unbreakable courage, unshrinking self-determination, and unquestionable driving skills. The men-folk knew from the beginning that this woman wasn't about to play by their rules, while at the same time, she'd stand for no preferential treatment because she was a woman. She was going to beat them just the way they had always beaten each other—and she did.

By now, her Hall of Fame racing exploits, four Top Fuel World Championships, and notoriously brassy verbal abilities are of legend. They are part and parcel of the Shirley Muldowney mystique, and there's little chance anyone would be curious enough to read this book unless they were already at least partially aware of her remarkable life and lore.

In 1995, I had the great pleasure to work with Shirley on a CBS Sports drag racing special called *The Ultra Team Challenge* at Old Bridge Township Raceway Park in Englishtown, New Jersey. Our friendship since then has blossomed, and throughout the remaining seven years of her driving career, we stayed in contact, occasionally crossing paths at NHRA national events. For someone who has always admired and respected what she did as both a racer and a pioneer, I truly feel privileged to have been given the opportunity to write this book, which puts to paper many of the dramatic, ironic, humorous, frightening, daunting, and tragic events in her tumultuous life.

Ron Lewis Photography

I sincerely hope the many tales Shirley shares in this book will not only serve as a source of great enjoyment and entertainment, but as a testament to what this great champion has come to symbolize in the world of motorsports. In true "Shirley" fashion, these tales are in her own words, told as only she can with her typical "no-holds-barred" style and signature "politically incorrect" use of language.

When we first began the process of compiling the tales you'll be reading in this book, Shirley was excited about the chance to share the many stories and experiences she has been a part of throughout her exceptional career. If you want a lot of "happy talk" or the watered-down versions of the countless stories that shape the life and times of this racing legend, you won't find them here. As always, she's telling it like it is.

That's Shirley Muldowney.

TALES FROM A
TOP FUEL DRAGSTER

Ron Lewis Photography

Chapter 1

In the Beginning...

Shirley Roque: Schenectady, New York

The somewhat unremarkable childhood of Shirley Muldowney would have never revealed to even the most prescient visionary of the day the future exploits for which this racing legend was destined. Although many great drag racers embarked on their quarter-mile saga thanks to the path already blazed by their parents, Shirley's formative years bore scant resemblance to what soaring adventures awaited her in adulthood.

Her family relocated from Burlington, Vermont, to Schenectady, New York, when she was just a youngster. Shirley shared a tiny bedroom with her sister, Linda, as their parents, Mae and Belgium Roque, huddled together with their daughters in a small apartment within the stifling confines of a four-story walkup.

Mae was a dedicated, hard-working mom who was the unbreakable mortar that held the family together, while Belgium—nicknamed Tex—was a hot-tempered, journeyman, country and western musician, laborer, and rough-and-tumble scalawag who spent more of his time gigging in clubs, keeping an eye out for a pretty woman, and constantly looking for the easy buck than doting over his wife and kids.

As Shirley moved into adolescence, she found waiting tables an accessible line of work for a young woman in Schenectady, and that was how she made ends meet. While still a teenager, she met and married Jack Muldowney, a local mechanic who shared Shirley's fascination with driving fast in the customized hot rods that were becoming more and more numerous in every corner of the country. Not long after their

My mom (right) was a strong, hard-working woman who held my family together. My sister, Linda (left), and I shared a bedroom growing up. *G. Harry Ransom*

marriage, the Muldowney family grew by one with the birth of their only child, John.

Overcoming the hardships of a blue-collar existence was anything but glamorous for a young wife and mother trapped in the economic quicksand that confronted many postwar families in the Northeast's industrial belt. It would be years before anything resembling prosperity would reward Shirley for her work ethic and determination, but those two aspects of her personality would pave the way for her later success.

Waiting tables and cobbling together the financial resources to enjoy the simple necessities of life were how she *spent many of her daytime hours while still a newlywed. But after dark, the lure of the rebellious lifestyle that street racing offered was an irresistible intoxicant for Shirley and Jack, and so it was on the late-night avenues and cross streets of Schenectady in the late 1950s that the seeds of Shirley's future racing legacy were sown.*

She Works Hard for the Money

My first waitressing job was at the old S.S. Kresge's 5-and-10 in Schenectady. I worked at the lunch counter that was right in the store, and it was huge. The counter began right at the front door and went all the way to the back of the store. It could seat about 50 people.

I had already left school when I worked there, and my friends would come in all the time and sit at the counter. I used to make them these giant ice cream sundaes whenever they came in. Really big sundaes. Lots of ice cream with the chocolate sauce and whipped cream falling off.

That got me fired.

It was fairly easy to find work as a waitress back then. I worked at a restaurant that had carhops for a while. I worked in the kitchen, and that's how I first met Jack. I didn't have a car of my own then, and he would drive me home. That was how we got together. But it wasn't until sometime later that I actually had my own car.

The first car that I could really say was mine was actually one that I found a couple of years after Jack and I were married. It was a 1940 Ford that was sitting in a dusty old barn. We paid $40 for it. After we bought it, Jack did it over for me.

Jack had spent his summers as a young boy at a place his parents owned up at Loon Lake in upstate New York near Lake George. He decided he wanted to leave Schenectady and take a job at a Ford dealer in Chestertown, which was about six miles from his parents' summer house. So we moved into a little apartment over a two-bay garage there and somehow survived together.

I got $10 a week for groceries and I remember going to the A&P where the farmers would walk up to the front door and kick the cow poop off their shoes before they walked in. This place was really out in the boonies, and it was quite a change for me.

During this time, my son, John, was born in Glens Falls, New York, and shortly afterward, we moved back to Schenectady. I found a job at a dairy downtown, adjacent to where the farmers would bring their milk, and those were the kinds of jobs I did to help us get by. Back then, you did whatever you needed to do to live.

Life on the Street

After I had gotten my learner's permit to drive, I was driving Jack's 1951 Mercury one night on Route 50, which ran between Schenectady and Lake George. Suddenly we looked over and a big, brand-new Oldsmobile pulled alongside of us, and for some reason, we started racing.

We were going pretty fast, around 100 miles an hour, and Jack reached over and said, "I'm just going to rest my hand on the wheel." We were on this long, sweeping curve doing anywhere from 80 to 100 miles an hour, and I remember him doing that. Was it because he didn't trust me driving that fast? I don't know.

We used to drive down to the city and look for drag races on State Street. I'd sit right next to him in his 1951 Mercury that had a column-shifted three-speed transmission and triple carburetors. We used to cruise around in it night after night, just the two of us.

Sometimes when we were stopped at a traffic light and a car would pull up beside us, Jack would hang out of the driver's window with both arms and start chattering with whoever was in that car—whether he knew them or not—and then rev up the engine.

When the light turned green he'd peel out with both of his arms out the window! He would work the gas and the clutch and I would steer and shift! We'd speed shift through the gears, and his arms would be hanging out the window the whole time. It was just a little game we liked to play to freak people out.

We were involved in a lot of street races back then. And I can remember some of the characters we used to run with in those days.

One was the son of a wealthy contractor. His name was Lou Tazzone. By then we had graduated up to a 1958 Chevrolet, but Lou always had a new Corvette. We had a 348-cubic inch V8, and his Corvettes had fuel injection. His Corvettes were always stock, and we were always trying to play "catch up" with him. Eventually we could get into Corvettes, but ours never had the fuel injection, because it was an extra option that we couldn't afford. We tried to get it to go faster with different gear ratios and things like that because they didn't cost as much.

Lou was a handsome "blond Italian," and he was a bit stuffy. He was one of those "Daddy, buy me a Corvette" kind of guys, and he always had sharp cars. We raced him on the street, and I remember beating him sometimes and losing to him sometimes. He didn't beat me because he was a better driver but because he had a faster car. But he really didn't like it when he was beaten by a girl. He hated me.

But street racing was a major part of our social activities back then. Jack and I weren't the type of couple who went on Sunday picnics. Drag racing was what we both loved doing, and we did a lot of it together.

I didn't have a lot of girlfriends in those days; I was a loner. I did have one close girlfriend, Linda Rubin, whose maiden name was Smith, and she's the only friend of mine I still keep in touch with back in Schenectady. She married a friend of mine, Bill Rubin, who was the only child of Dr. Israel Rubin, the doctor who devised the test for jaundice. In fact, it's called the "Billy Rubin Test."

Bill's father died, and then two weeks before his 18th birthday, his mother died, and I can remember hearing about the birthday party he had that lasted three days. Kids were staggering out of his basement for a couple of days after the party, and they were totally drunk. Bill wound up with a lot of money, but unfortunately, he pissed it all away.

He would go down to the Chevy dealer in town, buy two new Corvettes off the showroom floor, take them home, and cut open the rear wheelwells with a hacksaw so he would have enough clearance to fit big racing slicks. Bill actually owned a speed shop, which I think he opened just so he and his buddies would have a place to hang out. Sometimes he'd go down to the White Tower hamburger stand for lunch and order 200 hamburgers for the guys who were always hanging around with him. Like I said, he just blew his money like a drunken sailor.

Linda married him, and in those days, she was into street racing for a while. The last time she raced, she was driving her husband's coupe out on one of those country roads that ran through towns such as Voorheesville and Rensselaer, and as she pulled out to pass another car while she was wide open, the hood flew up and blocked the windshield!

And guess who was coming the other way?

Her husband!

Luckily, she didn't hit him, but what were the odds that he would be coming in the other direction at that very moment? That was the last time she street raced.

Officer Guy Barbari

Officer Guy Barbari of the Schenectady Police Department was the cop whom we all knew about and wanted to escape from if we were out racing. He was the cop who would come after us and chase us down at 100 miles an hour.

I can remember he'd pull Jack and me over in our 1958 Chevy, get out of his car, walk over to us holding his ticket pad, put his foot up on the bumper, and say, "Allright, what's the green light for?"

We'd answer, "That means, 'Go.'"

He say, "What's the red light for?" and we'd say, "That means, 'Stop.'"

Then he say, "What's the yellow light for, Shirley?"

I'd say, "That's to beat the red light, Guy!"

And he'd start writing.

That was what we did when we were kids. I had a smart mouth, and because of it, he wrote me up plenty of times. But for many years I exchanged Christmas cards with Guy and he lived well into his 80s. He died last year, and I still have the last Christmas card he sent to me.

"Real" Racing

Just about every social activity I was involved in had something to do with racing. On Wednesdays we'd all go to Fonda Raceway, and on Sundays we'd go to Glens Falls. Fonda was where I raced in the first organized event of any kind. It was an eighth-mile dragstrip that ran inside a half-mile dirt oval. They used to pack that place.

The first car I raced there was our 1958 Chevy before we bought our first Corvette. I won a lot of races back then, and I can tell you that the guys and the girls had a problem with that. The guys obviously didn't want a "chick" blowing their doors off, and the girls didn't like the

In 1963, Phil Castranova (right) and I get ready to race at the Fonda Raceway. *Simek Photography*

fact I was beating their boyfriends. That made it all the more satisfying for me.

I'd say that altogether Jack and I won about 200 trophies—little two-dollar trophies—after all was said and done, but I don't have any of them anymore. We sold them all back to the track because we wanted the money to buy a new lower gear rear end.

It doesn't bother me that I don't have those old trophies anymore. What would I do with them? And it doesn't bother me—at least I don't think it bothers me—that I don't have any of the cars I used to own. I did see my last Corvette about four or five years ago when I was matchracing at Lebanon Valley, New York, which was the last time I was there. The fellow who owned it had restored it, but I don't regret not owning it anymore. But it was fun racing it when I had it.

Jack

Jack took great pride in wrenching and preparing our cars, and I found that throughout the entire time I spent with him that he was a very unselfish man. A good guy. He had no problem stepping back and knowing what his place was, and that's not an easy thing for a lot of men to do. There were a lot of ways that we didn't mesh, but we're still friends. Since the divorce, I don't talk to Jack a lot, but I know where I can find him. I have a lot of respect for him. I think he's a great guy.

I kept Muldowney as my last name, because well, I guess you could say it's my trademark. People recognize it.

Ron Lewis Photography

Chapter 2

On the Strip

"Cha-Cha"

In *Heart Like a Wheel* the way the movie tells the story of how I got the nickname "Cha-Cha" isn't the way it really happened. Connie Kalitta didn't come up with "Cha-Cha."

Jack Muldowney claims that when my son, John, was born, one of our relatives came up with "Cha-Cha." I don't remember it happening that way. Of course, Jack has all kinds of stories, and that was just another one.

What I do remember is having "Cha-Cha" painted on the side of the old 1958 Chevy that Jack and I raced. It was written on the side with white shoe polish by someone at the racetrack where we were running that weekend.

Whoever did it was probably the person who used to write the numbers of the race cars on them as they were getting ready to race. And the fellow who did it had to be an artist of some kind because the way that he painted it on was cute. He hyphenated it and arranged the second "Cha" a little higher than the first "Cha," and we all looked at it and thought it was clever. In fact, we kept it on there until it was washed off by the rain, and then I had it painted on permanently. That's where my nickname began.

There are pictures of my old race cars, including some of my early dragsters that Don Long built, and you can see "Cha-Cha" painted on them. And that was long before I met Connie Kalitta.

Tommy Ivo, who was one of drag racing's greatest showmen, told me a long time ago, "Don't ever let go of that name and don't ever paint any of your race cars any

My race car in 1965 had the name "Cha-Cha" painted on it. Jack Muldowney (holding my helmet) worked on all of those early cars. *Shirley Muldowney Collection*

other color but pink." At the time I didn't know exactly what he meant, but as the years passed and I began to appreciate the importance of showmanship in drag racing, I understood his advice.

Some people think I fell out of love with the nickname "Cha-Cha," but it wasn't that I grew to hate it. It was just that I wanted people to know me more as Shirley Muldowney. There were people in high places and people in authority who would say, "Oh, sure, you're talking about 'Cha-Cha' Muldowney..." and nobody referred to me as Shirley. So that's why I began to back away from that whole "Cha-Cha" image.

Then Steve Evans stuck his two cents in. He was announcing on the PA system at a race and said, "Here comes Shirley 'Don't Call Me Cha-Cha' Muldowney," and that really took off! I kept reading that over and over and over again, and that's when I began to hate it. At that point, if I were standing in front of the grandstands and fans were calling out to me, if I heard "Hey, 'Cha-Cha'!" I wouldn't even turn around. They couldn't get a rise out of me.

Even to this day, people are always asking me to sign an autograph with "Cha-Cha" because they say it takes them back to a special time in their lives. I'll sign it that way for them, and I'm happy to do it.

Getting into a Dragster

Back in the early 1960s, I wanted to be part of the "Detroit scene," and by that I mean that the manufacturers were spending a lot of money on the big factory Super Stocks with the altered wheelbase setups, and they were sending teams around the country to race. For once, I was hoping I could get on the inside and be able to race without watching out for every nickel and dime.

I made a couple of trips to Detroit, I wrote letters, I did everything I could possibly do to attract their attention, but I couldn't get them interested in me. The people who were putting these teams together, such as Dick Maxwell at Chrysler, jumped on all those "California racer-types" and "corporate types," so finally I copped the attitude, "Screw the factory street cars, we'll build a dragster."

Well, Jack Muldowney gas-welded together a B/Gas dragster. It had an unblown Chevy engine with fuel injection, and the first time I ran it was at South Glens Falls, New York.

The dragster had an "up and down" pushbar in the back, meaning it was vertical, and it was right behind the fiberglass body. On our tow vehicle, which was a Buick Riviera, Jack had welded a horizontal pushbar that would meet the vertical bar in a criss-cross fashion. At each end of that horizontal bar, he had two probes that stuck straight out that would keep the vertical bar from sliding off when he pushed me. The first time he pulled up behind me to push-start the car, those two probes came right through the body and nearly impaled me! We hadn't thought about it! That could have been real bad.

I sat right over the rear end in that car, which was the way front-engined dragsters were built then. The rear end was really welded onto the chassis so it couldn't work loose and spin back into me. A lot of drivers got hurt that way back in those days.

After I made my first run in that dragster, I asked Jack, "How fast did I go?"

He said, "130."

I said, "I knew it."

Well, I went 80 miles an hour, but it *felt* like 130. Jack was messin' with me.

And the fans? They threw soda cans at me and some of them were full. They hated me. They were obviously jealous because I was doing something that they all wanted to do.

Super Stock Plymouth

There was this local fellow, Charlie Lendrum, who was pretty notorious in the Schenectady/Tri-Cities area at the time and who became a really good friend of mine. Later on, he worked with us on our cars and he was one of those guys who would get there and dive in, head first—a really gung-ho kind of guy, and we had to reel him in when he worked with us in the 1970s and 1980s. A few years ago he passed away, living alone, in a little motel. That was sad.

In 1963 I was driving a full-bodied, fully upholstered Plymouth with a 426 Max Wedge, and this car had a radio and a heater. It wasn't one of those stripped-down, superlightweight factory experimental cars that Chrysler was building then.

Charlie had one of those 1963 factory experimental Dodges with the aluminum front end and all the latest factory racing parts, and we were booked for a match race up at a dragstrip in Milton, Vermont. There were so many people there for that race they were hanging over the fence. It was the only time that my father came to watch me compete, and he was thrilled with the excitement and all the attention I received.

What was really funny was what happened when we arrived at the track. Jack and I and the fellow who owned that Plymouth were trailering it into the gate, and the track owner asked us how much we charged to run.

Jack turned around and called into the backseat, "Stan, what did we get for running at Atco Dragway?"

We had never even been to Atco in our lives!

But Stan answered, "Oh, I think we got $150."

So the track owner peeled the money right off and paid us upfront!

We agreed to a three-round match race with Charlie that day, and we won all three rounds. Stan was afraid to drive the car, which was why I raced it, and I still have the bracelet he gave me for winning that day. It's a silver charm bracelet with a charm on it that says, "11.78 E.T." That was the quickest Super Stock time ever, in fact, it was quicker than the record that Hayden Proffitt was sitting on at the time. And like I said, I ran that time in a fully equipped car, not one of those factory-prepared Dodges.

Charlie's car had a four-speed, and mine had a push-button Torque-Flite, and that was the difference. We won, and it really was a very exciting day.

Jack (right) works on the dragster before a race in 1968.
Shirley Muldowney Collection

Into the Woods

We raced the dragster some more and kept improving. But I'll never forget the night we were racing at Lebanon Valley. The car had cintered-metallic brakes, and Jack had somehow made a mistake when he adjusted them. He had backed them out too far so that after making my run, I had no brakes when I hit the finish line. I had the chute out, but it wasn't enough to get the car stopped. The car rolled through the shutdown area, through a fence, and into the woods. It was dark, I mean *really dark*, and I was just holding on, waiting for the impact. You couldn't see your hand in front of your face and how I avoided T-boning something is beyond me. I could hear the trees and branches whooshing past my head—*whoosh, whoosh*—until finally the car came to a stop. That could have been it.

We were able to tow the car out of the woods, and eventually we got it repaired to race again. I may have been shaken up for a quick moment, but the first thing that was going through my head was, "How are we gonna fix this?" I think that's what shook me up the most.

For a fleeting instant, it's possible that I wasn't sure if I wanted to get into that dragster again. But honestly, I was just a young kid, and I did stupid things!

When you're that age, you make decisions you'd never make as an adult. You're stupid! When you're young, you think you're going to live forever. I did, so I just went on with what we were doing and didn't think about it again. I wanted to race, and to me, that accident was just a single incident.

Muldowney and Kalitta

After Connie Kalitta took control of the Funny Car team, I moved from New York out to Michigan because I was looking for a change in my life. Jack wasn't happy about it, but his desires were not the same as mine as far as our future was concerned. That was the end of my marriage. I decided Connie was the guy for me, and we began dating.

The tough part was moving from New York to Detroit. But it was important to be in the center of everything that was going on at the time. Connie took it upon himself to book my car because he had all the contacts with the promoters. He hadn't been running many national events at that time so he depended on match-racing to keep him on the circuit.

Whenever a promoter called Connie would tell them that they couldn't book my car unless they booked his car, too. A two-car deal. It helped his match-race deals to have me included in the package.

One hand washed the other. He helped me; I helped him at a time when he really wasn't doing that much. We both raced Funny Cars in 1971, 1972, and 1973, and that's what brought in the money. In 1974 after all the fires in the Funny Car Connie had Logghe Stamping build a dragster for me that he paid for. The engine parts were mine, a lot of the other parts were mine, and just what Connie contributed then, I really don't remember. I do know that every cent that I made went back into our race program.

Shirley's First Funny Car

At the end of the 1970 season, I bought a used and whipped Funny Car from Connie, just the chassis and a Mustang body. Funny Cars were happening then, and that's why I wanted to make the switch.

In 1971, I began racing Funny Cars with Connie Kalitta. That year I won Rockingham Nationals; it was the first national event won by a woman. Jack (upper right) was there with me in the winner's circle. *Shirley Muldowney Collection/Paul Wasilewski Jr.—dragphotosinc.com*

I didn't have any trouble adjusting to the Funny Car after driving a dragster.

But the first time it caught fire at 200 miles an hour, let me tell you, I gained a lot of knowledge and a lot of experience real fast. I knew then exactly what I was riding in.

That first one I bought from Connie didn't have a butterfly-shaped wheel; it had a round one. And the front end had suspension with coil springs, but Connie chained the axle up against the chassis so it wouldn't pull the front wheels up when it launched and keep me from seeing where I was going. Eventually I ran quicker and faster in that car than Connie ever did.

The first time I raced it was at Lebanon Valley, New York. It was a night race, and everyone was there: Lew Arrington and his *Brutus* Pontiac, Leroy Goldstein and the *Ramchargers* Dodge, "Jungle Jim" Liberman, and a bunch more.

I won the race. I kicked their asses. And Kalitta was ecstatic!

When Goldstein towed back to his shop in Detroit, Connie was waiting for him and really gave him a good going over! Connie made sure everyone knew I had beaten them all in one of his old cars.

Funny Car Fires

The first time I had a fire in a Funny Car was at Orange County Dragway. There were a bunch of big names there but I remember Funny Car racer Paula Murphy's boyfriend was there, "Fat Jack" Bynum. I couldn't stand him. He was just a boisterous blowhard. In fact, the racers often joked amongst themselves that if someone didn't shape up, they should be sent to Fat Jack's Charm School. One time, I'm not sure when, but I happened to be riding in the truck with them while he was towing her race car. He was going around 90 miles an hour in the rain as if he was trying to impress me. Please!

I had made a run and the car blew up and started a fire. Bynum was at the end of the track and I got the fire out as the car was coming to a stop. I was trying to climb out through the side window of the car, and here was Bynum, with some help from some other idiots, trying to lift up the body so he could be a big hero!

I was screaming, "I'm coming out the window! I'm coming out the window!"

But he was showing everyone what a brave guy he was. There were a lot of guys like that back then.

The worst fire was in 1972 at Dragway 42 in West Salem, Ohio. I left the line with a hurt piston, and as I went down the track, it erupted in flames and engulfed the whole car. I had no brakes because the brake lines back then were rubber, and they got burned off. I hit the fire bottles, but the car went off the track into some rocks and boulders and just destroyed it.

That car was built by "Little" John Buttera, who was one of the most uncooperative chassis builders I'd ever dealt with—even worse than the Logghe brothers who built what I called *iron*. Heavy, crude, sloppy stuff. And they hardly ever had your chassis ready on time, unlike Al

This 1973 fire at Indy melted my goggles and burned my eyelashes off. *Shirley Muldowney Collection*

Swindahl, who was a great chassis builder who always treated the racers right.

But Buttera's car was bad. Not everywhere, but in some places he just didn't pay attention. It had a nice yellow paint job, but I can remember running that car and when I got down near the finish line at 200 miles an hour, the body and the tinwork, which the body was mounted on would separate, and I could see two inches of daylight between them. Buttera didn't mount the wheels in such a way that they were centered in the wheelwells. It was really noticeable, and when we mentioned it to him, he just ignored us. Connie let him get away with it, and I was a little surprised. I even told Connie that if I ever had a fire in that car, it would be a bad one, and sure enough, not long thereafter I had that one at Dragway 42 and got burned.

Buttera could be rude and arrogant. He was the least impressive of all the chassis builders I worked with. He had no respect for me. After that fire that I had in his car, he was quoted as saying while he was sitting in his body shop in California that it was my fault because I didn't know how to drive.

Buttera had relatives who lived near Union Grove Raceway in Wisconsin. In later years when I raced there, Buttera would come to the track and say hello to me. He couldn't have been any nicer, and it just goes to prove that people can change.

Shirley Burns Again

I went to Indy in 1973, but Connie wasn't with me. He was off doing something else, I don't remember what, and Poncho Rendon was running the car with a bunch of volunteers. I made a run, and just as I was hitting the traps and lifted off the throttle, the engine exploded.

Let me tell you, it was so hot. The fire was streaming out the side windows, and the heat was intense. There was no oxygen. I still remember the feeling of that fire surrounding me. The fire crew was actually pulling alongside of me as I was rolling to a stop, hosing the car and yelling to me, "Keep it straight! Keep it straight! Now pull to the left!"

Don Prudhomme was right there, and he tells the story of how I was worried about how my eyelashes looked, but actually, I was kidding. My eyelashes were just about burned off! It burned me like the fire I had in 1972 at Dragway 42. My goggles were burned off.

I prepare for a timed run in Poncho Rendon's Top Fuel car. This run helped me get my Top Fuel license and was my first pass in the Top Fuel class. *Shirley Muldowney Collection*

The area around my eyes was burned pretty badly. Not my eyes themselves, but all around them. But believe it or not, a week later I had a match race in Toronto against Ed McCulloch, and I was there with bandages on my burns. We repaired the body on the car and after fiberglassing it back together, it weighed another 200 pounds. But we made it to that match race, and we beat McCulloch! And he didn't like it!

Sitting up Straight

The sensation of acceleration and speed never bothered me. I moved up gradually and always understood my limitations. The cars and equipment we have now are nothing like what we had back then, but one thing I always knew right from the get-go was that I had to sit up

Throughout the early years of my Top Fuel career, my son, John, was a hard-working member of my crew. Here he is helping me as I prepare to fire up my rear-engined Top Fuel dragster. *National Hot Rod Association/National Dragster*

straight in the car. There was none of this, "Let's be cool and laid back in the car." I knew that wasn't going to work for me. You couldn't hang on to your equilibrium.

When I first drove in a Top Fuel dragster, Poncho Rendon's car in Cayuga, and stepped on the throttle, my head was shoved back and I couldn't collect myself so we took a bunch of shop rags, rolled them up, and tucked them behind my head. I was always adamant from that time on that my cars had to have that headrest. There were some racers who got into Top Fuel cars and never picked up on the headrest. That was their only problem. They weren't sitting up straight.

Now they have that G-strap that connects from the front of the helmet and holds your head down, but I never used one. I hated them. I hated throttle-stops, which limit the throttle travel on a race car during the burnout and prevent the driver from accidentally over-revving the engine. I wanted to race the way that was most comfortable for me.

"Hot" Magazine Layouts

The hot-rodding magazines have always been good to me. Back in the Funny Car days and the early Top Fuel days, I appeared in some photo layouts with hot pants and miniskirts. I can remember at one shoot the photographer, Steve Reyes, had his girlfriend with him, and I actually wore the outfit she showed up in. Hey, why not? Her outfit looked great in the layout.

Back in those days, there was nothing wrong in trying to get a little "ink," and that's the way we did it. I even had an ABC Sports producer come to my house and take a look at some of my old photographs, and she couldn't believe the eyelashes I had.

"Shirley, are those fake eyelashes?" she said.

"Of course," I told her. They were artificial because my natural ones had been burned off in a Funny Car! So, yeah, I wore artificial ones, and she picked up on it right away.

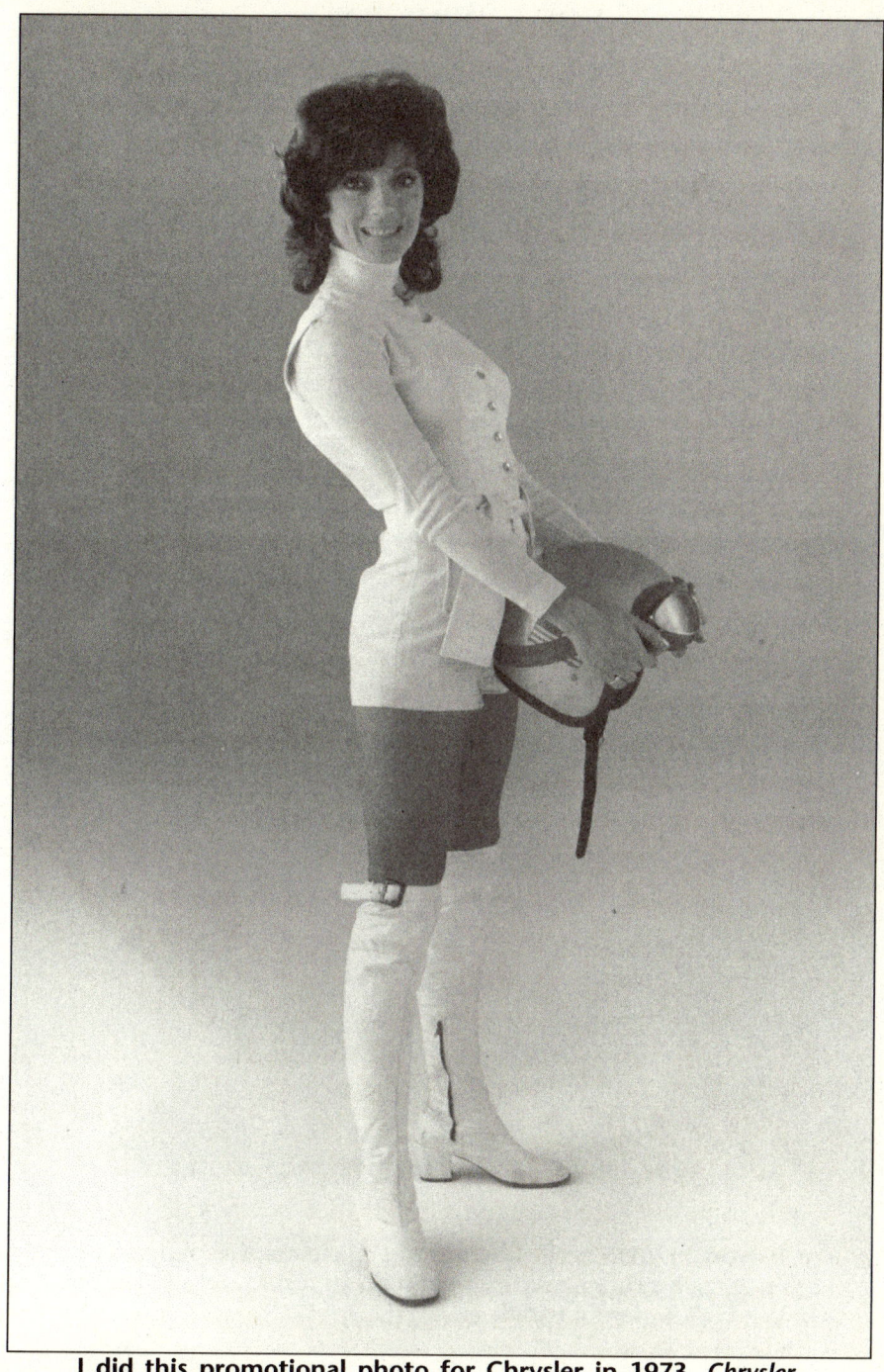

I did this promotional photo for Chrysler in 1973. *Chrysler Historic Archives*

Changes

A lot of the changes that were made in drag racing over the years were good ones. Now I think it's good that there are throttle stops. I believe Lee Beard was the first one to install a throttle stop. I think I was the first driver to have plexiglass windows built into the firewall of a Funny Car body. It allowed the driver to see the engine in case of a fire. We put them into my car after the first time I had a fire in it, and then everyone had them.

At Indy in 1975, Connie Kalitta (right) gave me this pep talk prior to facing Don Garlits in the finals. Unfortunately, I lost.
Shirley Muldowney Collection

Columbus, 1976

That was quite a weekend. Remember, in those days we ran five rounds in eliminations because we had 32-car fields. It was a long day to get through all those other cars and get the win, and I had been runner-up there in 1975. But my mother was in Columbus in 1976 to watch me race, and it was special to get my first win with her there to see it. I beat Bob Edwards in the final, another racer who is no longer with us.

I scored my first national Top Fuel victory at the 1976 NHRA Springnationals in Columbus, Ohio. Included in the victory celebration were (from right to left) NHRA president Wally Parks; my son, John; an unidentified model; Connie Kalitta; me; and Connie's son, Scott (far left). *Shirley Muldowney Collection*

That race was a long time ago, and I don't remember a lot about it, but I remember Bob Daniels, the NHRA's Division 3 director, was elated. I have pictures of him after I won, and you can see he is really excited. Ten years earlier, he and his wife, Eileen, weren't very anxious to see me out there racing, but after a while, people change. Like I've said, once they saw that I could put people in the grandstands, it had a lot to do with changing things.

1977

I had that first championship sewn up at Indy. There were nine races on the schedule back then, and I went into Indy having won the last three events in Columbus, Englishtown, and Montreal. But back then, there were points races, or divisional races, which we don't do anymore, but there was a lot of racing going on to win the championship.

I don't remember how we did at Indy that weekend, but I remember that when we had actually clinched the championship, Linda Vaughn came over and gave me a big hug.

She seemed to be as delighted as anyone. The fans went crazy when the announcement was made that we had won the championship, but I'll tell you when they really went crazy, and that's when we won Indy in 1982.

Englishtown, 1977: My crew and I celebrate a win that led to my first Top Fuel championship. *Shirley Muldowney Collection*

Different Lifestyles

When it was over between Connie and me, it was just a case of our lifestyles becoming different. It just wasn't working for me, and I said, "I'm over this."

Connie and I got into it during a weekend in which we had a match race at Orange County on Saturday night

My first championship (from left to right): Scott Kalitta; Rahn Tobler; Miss Winston; my son, John; me; Wally Parks; and Connie Kalitta pose for a photo at the Winston Award Ceremonies. *Shirley Muldowney Collection*

and one in Fremont on Sunday. The car was being hauled by transporter while Connie and I were going to fly. Before we left, we got into a big argument at the motel we were staying in, and it was intense. When we got to the airplane, we didn't speak to each other at all for the entire trip. When we landed, I got out of the plane and never even picked up my luggage.

As I walked toward the terminal, Connie called, "Shirley! Shirley!"

But I never turned around, got into the first taxi I saw, and left him standing there at the airport. It was over.

When I drove to Fremont, I said to Rahn Tobler, "This is your deal if you want it."

And he said he did. I hired him to come work for me as my crew chief. That was the end of it for Connie and me, but it was soon the beginning of things for Rahn and me. After years of running my crew, we were married in 1988.

Poor Don Lampus

Sometimes I think the fans can go a little overboard. I was racing Don Lampus, a former NHRA Rookie of the Year and a very friendly guy, at an event in Norwalk, Ohio. When Don beat me, a woman got so mad that she went over to him and physically attacked him! He laughed it off, but is that funny?

"Warfare"

After I broke away from Connie at the end of 1977, he wasn't about to let things end without making his point.

We fought back and forth a lot while we were out there racing separately. He would loan motors or anything else he could to whomever we were racing to help beat us. He'd book where we were booked in, chase us down, and in general, really become a nuisance. Rahn Tobler, my crew chief, always supported me as did my son, John. But I've called it war, and that's what it was when it came to racing against Connie.

After Connie and I split in 1977, every time we raced each other seemed like a brutal battle. *Auto Imagery Inc.*

It was warfare.

One year we were at Englishtown and Connie got by me in eliminations. As I was coming back down the return road, his truck went by and I sailed my helmet right into the side of it! He saw what I did.

"Look what you did to my truck!" he yelled.

He never came after me, he was pretty upset by it, but he was laughing at the same time.

Witnessing Disaster

Whenever I'm asked if I've ever seen a nasty accident, I think back to Indy in 1979. A bunch of "weekend warriors" put enough money together to bring a Top

Fuel car to the U.S. Nationals and a guy named Frank Rupert was driving it.

I was up in the tower and had a pretty clear vantage point of his crash that hurt him pretty badly, but the worst part was a TV cameraman, Joe Rooks, was hit by Rupert's blower and was killed. He was panning as the wreckage was flying down the track and this 100-pound killer hit him from behind. He never knew what hit him.

The World Finals, 1980

We went to the World Finals in Pomona in 1980 with a shot at my second championship. There were three other drivers who were also in position to win the championship that year. I qualified ninth, and in those days, we had 32-car fields, which meant I raced the No. 1 qualifier in the first round. I won that round and that put me into the second round to face Connie, and I beat him, too.

At that point, Connie was out to help whomever I was going to race that day. He was going into his "I'll loan you a motor" routine with my opponents, which was what we had seen him do before. "I'll give you a set of tires! We've got to stop her!" and all the usual offers he'd make to help us get beaten.

My mother was there that weekend, and it was the first time she had ever flown on a big airliner. She flew all the way from Schenectady with Charlie Lendrum.

That weekend at the 1980 Finals, I made it to the final round to race Frank Bradley. Frank wasn't someone I knew personally, but I knew he had said some very unkind things about me back then. I was told it was he who had hung a water-filled prophylactic on the back of my trailer in the parking lot in Columbus. I really didn't like him, even though I didn't really know him. I'm aware of a lot of things like that that he did that were really mean. And you don't want to get into that "girly' thing with me. He did, and I don't like that.

You just don't do things like that. It's not cool. There's not a racer out there who would like it if my husband pulled that stuff on their wives, right?

But as we were pulling into the lanes to race Bradley in that final, I pushed the clutch pedal to the floor, and it stayed there! We didn't know what the problem was, and as the guys were trying to get it fixed, Steve Gibbs, who was the director of racing for the NHRA back then, came over and said, "Don't worry, keep working on it. Bradley will have to wait."

Finally, we were able to get the clutch straightened out and we absolutely freight-trained him. I won the championship in 1980, and I can't begin to tell you how good it felt!

The AHRA-Garlits Conspiracy

There were many examples over the years of favoritism that some drivers would enjoy because of who they were. Don Garlits is one of them, and this is what happened one year when we were running an American Hot Rod Association event in San Antonio back around 1981.

Don Garlits and I talk away from the dragstrip. *Shirley Muldowney Collection*

It was a very hot day, around 105 degrees. Everyone was feeling the heat that afternoon, but we raced anyway. Garlits was racing Jody Smart in one of the earlier rounds of eliminations, and Jody got a redlight. Word was that the redlight had been purposely tripped by someone in the tower under the orders of the AHRA's owner, Jim Tice. Everyone saw what happened. There was no question they gave Jody that redlight by design. It was obvious.

Jody was furious. He was one of the nicest, mild-mannered people you'd ever want to meet in your life. He was a quality gentleman. He charged over to the track office, which was just a trailer parked behind the popcorn stand where they counted the ticket money and hid it in popcorn bags so they could get it out of the track without being mugged.

When Jody got to the office, Garlits was there with Tice.

Jody looked at "Big Daddy" and said right to his face, "I'm going to tell you something. All of my life I've admired you. When I was a kid, I looked up to you, and now, I wouldn't walk to the mailbox for you! You're a no-good cheater, and I'll never forget this!"

Jody walked out of the office and headed right for the starting line. When he got there, he took his bare hands and began trashing the Christmas tree! He attacked it, bashed it, kicked it over, beat it into the ground, and did a real number on it. There were 10,000 rowdy drag racing fans in the grandstands, who saw all this and went crazy! San Antonio was Jody's hometown, and so the crowd was lovin' what they were seeing.

Well, that shut the event down on the spot. They had no Christmas tree and no timers, and everything came to a halt. Before long the fans began to get a little antsy, because the race had been stopped. They were partying to be sure, but they were getting impatient, too.

All of the racers were milling around at the starting line waiting to see what was going to happen as the track officials were trying to sort things out. One of those racers was Don Prudhomme. Suddenly, an empty beer can was tossed out of the grandstands and landed right near where we were waiting. Prudhomme did the worst thing he could have done.

He picked up that beer can and threw it back into the crowd.

I'll tell you what happened next. About one thousand beer cans came out of those grandstands in our direction! There were beer cans in every direction all over that racetrack! I can't tell you how long it took for them to clean up all of those cans!

Eventually things got straightened out, and by the time the final round was set to go, it was about 1 a.m. I was going to race Garlits. And I won.

It was great. It was really great. After I stopped at the end of the track and turned around to come back, the plan was for me to let out the clutch as we were pushing back and restart the engine just to give the fans a little extra jolt. Rahn Tobler, my crew chief, decided not to, though, because he wasn't sure if we had enough fuel left in the tank.

But that was only one example of how Garlits and the AHRA worked together. Let me put it this way: Garlits had the AHRA on his side. He was an avid AHRA supporter. When I won the AHRA World Championship in 1981, Garlits was not happy about it.

And that made it all the more satisfying for me.

Styx

It's funny how we got to know the band Styx. We were at a big invitational race in Union Grove, Wisconsin, in 1980. We were down in the pits, and John was under the car, working on the bottom end, and I was standing beside the car, probably smoking a cigarette. There was a big crowd down there, and we were all milling around before we had to go race.

I looked over at the crowd, and there was a face that stood out, so I leaned down to John and said, "John, look out at the ropes."

He looked over there and said, "Wow! It's James Young!"

James Young, or "JY" as he's called, was standing there with Chuck Panozzo, who was the bass player and his brother John, the drummer.

Oh my God!

We came out, brought them into the pits with us, and they hung around with us for the rest of the day. We went out later and had dinner with them, and it was the start of a wonderful relationship that has lasted until this very day.

We got to go to quite a few concerts as their guests in so many different places. They were on the set of *Heart Like a Wheel* when we were filming it. We got to spend some time with them when they were shooting the music DVD "Mr. Roboto," which is one of my favorites.

We took the *Paradise Theater* album cover, which is a beautiful piece of art with a lot of detail, and had John Pugh hand-paint it on the cowl of the race car. And let me just mention that John is an incredible artist, who has done the fantastic paintwork on all my helmets too.

Anyway, at the end of the 1981 season, we gave that *Paradise Theater* cowl to JY. In 1982, he was standing at the starting line when I won the U.S. Nationals in Indy. And if you have a copy of their *Paradise Theater* album, look on the back of the cover and you'll see, "Thanks to Rahn and Shirley."

Indy 1981

The U.S. Nationals in 1981 is kind of a tough story. That year, the people who were working on the *Heart Like a Wheel* movie were at Indianapolis shooting scenes for the film. But prior to that race, there were teams that were having horrible problems with pistons coming apart. The pistons were breaking up under power and disintegrating in the cylinders. It all boiled down to this one company, JE Pistons, that was supplying everyone at that time, and those pistons weren't forged; they were cast.

We wound up blowing three engines at Indy, and that just about put me out of business. Keith Black, who was one of the greatest engine builders in history, did something I never asked for, and that was extend some credit to me so I could get the parts I needed to run. He did that and I survived that mess.

I heard at the race in Denver, which we did not attend, most of the Top Fuel cars with the JE Pistons were having that same problem. Engines were exploding, and it was all because of those inferior pistons.

The owner of this piston company at one point came to my crew chief, Rahn Tobler, and told him, "You don't know what you're doing." That made Rahn really hot.

The following year, 1982, we changed from those pistons, as did a lot of people, and although I didn't win Pomona because we were having a problem with the return lines in the fuel system, we went to Gainesville and won there. And as soon as my car crossed the finish line, Rahn tracked down the owner of JE Pistons and told him to his face, "Don't you ever tell me again that I don't know what I'm doing!"

That fellow had manufactured an inferior product, and he wasn't man enough to step up and do something about it. And I can tell you that when we switched to Venolia pistons, the brand I used throughout the rest of my career, we never had another problem like we had with the previous ones.

I can remember a while after that problem I was walking around in the pits at a race somewhere and I was offering some people some chocolate chip cookies that I had. Well, that owner was standing nearby and I offered him a cookie.

"I don't know if I should eat one of those," he said half-jokingly.

Well, at least he knew where we stood.

Don't Be Predictable

I can remember in 1982 when I was running for my third championship, I watched an interview on television where Mark Oswald, who was a great guy and a very good driver for Candies & Hughes, was asked what he thought about this driver, that driver, and finally what he thought about Shirley Muldowney.

He said, "Shirley's a great driver, and I know exactly what she does and how she does it. She pulls up, and she waits a little bit and then cleans the motor out before she rolls it all the way in."

I never forgot that.

The next time I ran him—I believe it was in Brainerd, Minnesota—I pulled up and did what I usually did, but when I snapped the throttle to clean out the motor, I did it twice. Well, that threw him completely off because I don't remember if he left too soon and got the redlight or was dead late, but I got there first.

Columbus, 1982

In 1982 in Columbus, I raced Lucille Lee in the final round, but there's a lot more to the story than that. Before she ever had to race me, she had to race Jeb Allen, who threw the race. I'll explain how we knew that.

Allen said his engine was broken and he was afraid to run it, so that gave Lucille a single into the next round. By then she had blown two of her engines, and so heading into that next round, her guys took the engine out of Allen's car—the one he said he was afraid to race with—and put it in her car!

The drama was unreal! There was almost a fistfight down in the pits. Connie was involved in it; Ronnie Capps

I raced Lucille Lee in the 1982 finals at Columbus. *Auto Imagery Inc.*

and Dale Armstrong put in their two cents. In the midst of all this, Allen claimed that Mark Danekas had loaned him that engine in the first place so he was obliged to give it back, but we knew Danekas had paid him off.

Whatever was the case, Lucille got to the final.

And I drilled her.

I beat her off the line, and I ran low elapsed time and top speed of the race. There's a video I still have that was taken from the tower, behind the starting line that shows the e.t. and speed coming up on the scoreboard as my winlight came on. Then you can see Pat Galvin run from the right side to the left side of the racetrack, face Danekas, and gesture with both arms, *"Take that!"*

Man, it was a great day!

After the 1982 Indy Win

As I look back on my career I can honestly say I would trade my three NHRA championships for one win at the U.S. Nationals. That's how big a race that is to me. When I won it in 1982, the crowd went absolutely nuts. Remember, I beat Connie in the final round five years after we had broken up, and it was the quickest and fastest side-by-side race in NHRA history at the time, and a woman had never won Indy before.

Connie was an absolute peach at the end of the track after we got out of our cars. He took it like a total gentleman. And the fans were wild. Just wild. I'll always remember that.

My crew—Pat Galvin; Francis Gray; Rahn Tobler; my son, John; and Charles Lendrum—and I celebrate winning U.S. Nationals in 1982. *Auto Imagery Inc.*

Rahn and I celebrate my 1983 win at Pomona. *Shirley Muldowney Collection*

"The Accident"

In July 1984, Shirley was seriously injured during Top Fuel qualifying at the 14th Molson LeGrand Nationals at Sanair International Dragstrip in Montreal. The debilitating injuries to both of her legs led to a long, painful recovery period, and to this day cause some lingering pain and lack of total mobility. To her many fans, the incident has become known simply as "The Accident," and she recalls the crash and its aftermath.

In 1984 I was in a terrible, debilitating accident in Montreal. I severely injured my feet and legs, and the car was completely destroyed.
Shirley Muldowney Collection

I remember that day. It was a Friday, and I wore my Trick jeans and pink cowboy boots to the track. Between the first and second qualifying passes that day, I went over to Al Segrini's truck and sat down for a little while and had a ham and cheese sandwich he had offered me. Just one of those little things that sticks in your mind.

I went back to my pit and got ready for the next qualifying run. I went to the line to make the pass and Dick Lahaie was in the other lane. I was in the left lane, and I remember I left before Dick did. Everything seemed fine during the run, but just before the finish line, I could see one of the front tires was losing the inner tube and the front end became very heavy, and then heavier, and then heavier still—making it very hard to steer.

My thought at that moment was, "Do I take my hand off the steering wheel to pull the parachutes?"

I pulled the cord for the parachutes and put my hand back on the steering wheel just as the front wheels jerked over into a hard left, at almost a 90-degree angle. I remember hitting twice, hard. What the car did was hit a culvert next to the track and then hit the embankment on the other side of it. I hit that embankment head on, and the car just disintegrated. The car literally came apart.

The first injury I sustained happened before the car took that big hit. When the front end made that quick jerk, the steering wheel whipped around and severed my thumb. It was just hanging by a piece of skin, and I was lucky that the doctors were able to sew it back on. You can still see the scar today and there is a difference in the angle of that thumb compared to my other one.

My crew chief, Rahn Tobler, pretty much took control of the situation at the end of the racetrack. But when they got down there, they couldn't find me. I repeat: *They could not find me!*

When they got there, there was no semblance of a race car. Finally someone started waving to the rescue personnel, pointing to the spot where I was, which was about 300 feet from where I went off the track. I was still strapped in the rollcage, which had broken off from the rest of the car right where the top of my thighs were. I had slammed head on into a muddy embankment and I tumbled another 300 feet finally coming to rest in an open area. What was left of the car and I were both the same muddy color, which was why they couldn't find me at first.

I was lying on my side. They got the rollcage back upright, which was about the time I regained consciousness. I remember thinking when I came to that I was out of the car, because I didn't see any of the car in front of me. But as soon as I woke up, I was in incredible pain.

I asked Ronnie Davis of the Safety Safari to give me something for the pain.

He said, "I can't."

But if it were not for Ronnie, I probably wouldn't have my feet today. (Some time after that, when the NHRA let him go, I was so furiously mad. I felt they were robbing the racers of somebody who they needed in case of an emergency.)

And I'll never forget "Diamond Jim" Annin who knelt next to me the entire time as I sat in the rollcage and held my head stationary. He stayed right there the whole time.

(Jim and I were good friends right up until his death in 2004.)

But my leg injuries were open and extensive. They were covered with dirt, gravel, stones, grass, mud, and oil. The medical people were trying to clean me up as we headed for the hospital. Then the ambulance got stuck in the parking lot as it was driving across the muddy field to get to me. What was really amazing was the next day there were hundreds of cars parked right in that field where I had gone off the track. No guardrails, no fence. Just that culvert.

As the ambulance was preparing to take me away, Rahn made sure there was someone who spoke French coming along with us so we would be able to communicate during the trip. He wanted to make sure he knew what the hell was going on.

It took 45 minutes for us to get to St. Hyacenta's Hospital and for them to stabilize me, but nobody there spoke English. I remember putting up a pretty big fuss because I was in a lot of pain. From there, I was taken to Montreal, and on the way, Rahn asked the ambulance driver to turn off the siren because it was upsetting me. Like I said, I was in some real pain.

When I got to the hospital in Montreal, they looked me over, X-rayed me, and gave me an arterialgram, which was very painful. Dr. Larry Conochie was called at his home to come in and see me.

His wife, who is also a doctor, asked, "Why are you going in to see a woman race car driver?!"

But she and her husband became very good friends of ours and still are after all these years. They're both wonderful people. And I'm so grateful to have had Dr. Conochie there when I was hurt.

Two months later, Connie flew me out of Montreal. And it was a horrible two months. As horrible as it could have been. I lost a lot of weight. The pain was just indescribable.

It didn't occur to me at the time that I wouldn't race again, but Rahn wanted nothing to do with drag racing. The immediate concern was that I might never walk again. They just were not sure. But the decision as to whether I'd race again didn't come for quite a while into my recovery. As my recovery began, I was sitting at home trying to figure out how we could dig our way out of where we were. None of us had jobs.

An Outpouring of Respect

After my accident, the other racers were great. They were terrific. I still have the pad of paper that Rahn used to take down messages from the racers, fans, and friends while I was in Montreal General. I still have those notepads.

I even got one message from my Top Fuel adversary Richard Tharp who said, "You tell her she's too goddamn mean to be lying in a hospital bed!"

Steve Earwood, who was the NHRA's public relations director at the time, set up a separate phone line in the hospital to take phone calls coming in because the hospital switchboard couldn't handle them all. There were 500 flower arrangements that arrived that they had to give out to nursing homes. It was out of control.

(The day before my accident, Steve and I had made a promotional appearance in Burlington, Vermont—the town where I was born—and we drove over the border to Canada to get to the race. Steve wasn't at the track when I crashed because he was out doing publicity for the event. But he was a very good friend, and one of the biggest mistakes the NHRA ever made was to cut Steve loose. The second biggest mistake they made was letting Dave Densmore go. Two of the worst things they ever did.)

A Less Than Perfect Recovery

My recovery was rough, and I ran into some trouble with a doctor who was treating me at Detroit Receiving Hospital. He was the head of orthopedics there, and at one point, Rahn went nose to nose with him. This guy was a joke.

He told Rahn one day, "You race the cars, I'll fix the bones."

And Rahn was *that far* from his nose. *(She raises her hands six inches apart.)* This was about seven months after the accident, and I was sitting on the examination table so the doctor could cut the cast off my leg that he had put on. He had no business removing the fixators on my leg, but he did. Then he told me to lift my leg, and when I did, my

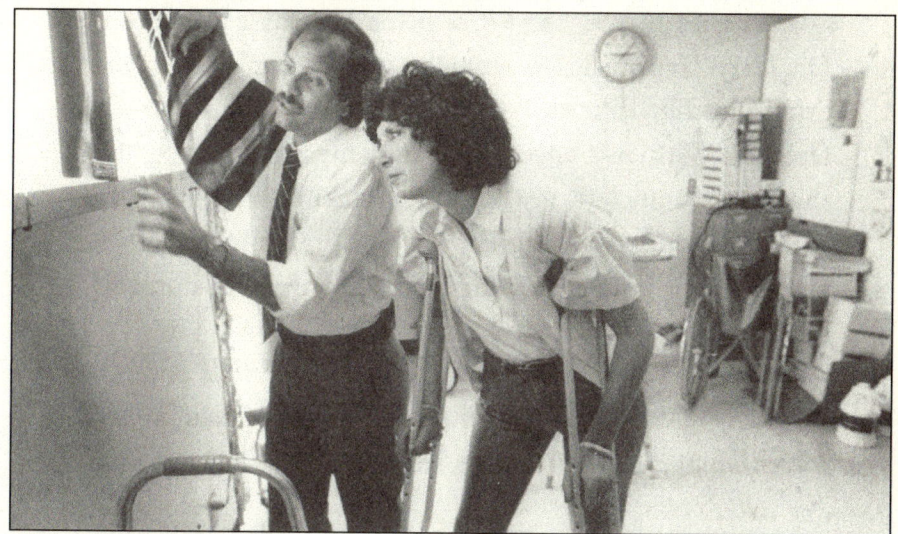

Terry Trammell goes over some X-rays with me during my recovery in 1985. *Taro Yamasaki/People*

heel stayed on the table and my leg pivoted like it was a hinge! Rahn went nuts!

This doctor denied me adequate pain medication. In my opinion, he was bad. I thought he was an idiot.

When Connie flew me down from Montreal, I never even saw my house. They transported me by ambulance right to Detroit Receiving Hospital, and we had been referred to a plastic surgeon in Detroit by the terrific doctors I had had in Montreal. But here was this incompetent head of orthopedics standing there when I arrived, and I was put in his care.

The very next day, he sent me into surgery and that's when he removed the fixators on my leg. My leg literally fell apart on the operating table. Then he fitted it with a cast while the bones were misaligned. And that made my recovery much longer than it should have been.

Thanks to Dr. Larry Conochie and my good friend, Dr. Terry Trammell, I was able to walk again. There is one thing that Dr. Trammell wishes could have been done differently. With my fused ankle and my right leg slightly shorter than my left, my body's alignment isn't quite right, and that's caused me some pain and discomfort all these years. He wishes that the surgery could have been done differently. So do I.

On December 15, 2004, Dr. Trammell and his team surgically replaced Shirley's left knee. The surgery was deemed successful, and she continues to recover from the aftereffects of her 1984 crash.

My fused left ankle created a real challenge for Dr. Trammell, but after this last operation, I am doing much better!

The Comeback

By the time I was ready to go racing again, I had been in therapy in Northridge, California. I was going to water therapy every day except Sunday, and it was a long, strenuous process. But I was raring to go, and that helped me get through that therapy from September to January as the first race in Pomona approached in 1986.

My first race was actually in Phoenix at Firebird Raceway, the track that Charlie Allen owns, a couple of weeks before Pomona. He had the biggest crowd in the history of his track when we showed up. A lot of racers didn't like Charlie, but we were always friends.

He is a tough but successful businessman and has nicknames like "Cheap Charlie" and "Save A Buck Chuck," but I always got along with him. He always treated me well.

I only made a couple of runs that weekend because there was a problem with the car—I don't recall what it was—but Rahn decided we shouldn't take a chance. He didn't want to take any chances and neither did I.

While the dragster for the event in Phoenix, I climbed into the car back at our shop and said, "I can't drive this car."

Rahn asked me, "Why not?"

I told him the clutch pedal was a problem.

The clutch pedals in a dragster are mounted with a slight angle where your foot rests on it. With my fused ankle, I couldn't pivot my foot to push down directly on the pedal as I normally would. My foot was hitting the bar that the pedal was mounted on.

I stressed to Rahn, "I'm telling you, I can't drive the car!"

My son, John, worked late into the night fashioning a new pedal for me that was straight up and down and with a big pedal surface, so I could get my foot squarely onto it. It worked, and that's the kind of pedal I had in every one of my cars until I retired in 2003.

It was perfect for me.

Doug Herbert to the Rescue

Jim Bailey started playing games with me at the starting line not too many years ago. I was racing Bailey one time, and he wouldn't stage, so I just sat there giving it right back to him. I heard from somebody that he said something like, "That'll be the last time she ever does that again."

Doug Herbert found out about it and went over to Bailey.

He told him, "I'll kick your ass if you ever mess with her again!"

What was funny about that whole episode?

That was when Bailey was driving a second car for Herbert.

A Close Call

During qualifying for the Winternationals in Pomona in 1989, Eddie Hill had his blowover. Just as he was getting to the finish line, his wing broke and that sent the front end up like an airplane taking off. We were on the return road right next to the track towing back, and Rahn was in the car. My first thought was, "Oh, my God! Don't let Eddie hit us!" Luckily, he didn't, and Eddie wasn't hurt. In fact, he came back and raced on Sunday.

Doug Herbert and I have been friends even when we have had to race one another. *BME Photography*

Improper Pricing

In 1992, we towed up to Edmonton in Canada for a match race, and I have to be honest, I never saw so many people in my life! You couldn't have squeezed another fanny into another seat in that place. When we pulled out to the starting line before the race began, they cheered as if the race had just ended! The Canadian people are absolutely crazy about drag racing.

At the end of the day, the promoter, who had been up in the big covered grandstands sitting with all of the VIPs, came over to me, and when he extended his hand to shake mine, the first thing out of his mouth was, "I don't think I charged enough money today."

Japan

In 1993, I went to Japan to race with a lot of other NHRA drag racers, and it was quite an experience. Craig Treble, who races the Pro Stock motorcycles was there; Randy Goodwin was there—quite a group of us all together. You can ask anyone who was along for that trip and had been on any of the other trips to Japan that the NHRA had made over the years: Ours was the most fun, the smoothest, and best organized one they ever had.

The race cars were loaded into containers down at the shipping docks in Long Beach, California, and the NHRA requested that we all be there on a certain day at a certain time for everything to be done properly. We had to basically map out the containers that our cars and equipment would be shipped in. I can remember how everyone stood around in sheer amazement watching Rahn put this whole thing together.

Melvyn Record of the NHRA saw what Rahn was doing.

He yelled, "Stop! Everyone drop what you're doing, come over here, look at this, and learn something!"

Rahn was just the master of organization. He still is. It's amazing.

Two weeks later we were in Japan, but even though we did everything we could to get the race in, our first race was rained out in Oita. When we were in Oita, the man who had financed the Japanese side of the tour came with his wife to see us and he told Rahn through his interpreter, "I know that when you get to Mt. Fuji, I think you will make the perfect run."

And Rahn shook his hand and said to him, "Yes, in Mt. Fuji we will make the perfect run."

With the rainout, we had a week to relax and travel around in Tokyo and the surrounding areas before we arrived in Mt. Fuji. It had rained earlier in the day, and they made an attempt to dry it, but for the most part it dried up on its own. That track was like a skating rink. Rahn knew we didn't have any kind of a racetrack to work with, so he took some power out of the engine when he tuned it up.

The fans didn't seem to mind. One thing you have to know about fans in Japan is how reserved and unresponsive they usually are. You can't always tell when something turns them on because they aren't very expressive.

Well, I did my burnout, backed the car up, rolled up to the starting line, and reached for the handle that turns on both fuel pumps. It's what we call "going to the highside." As I did that, the handle broke off in my hand! It's not something that happens very often, but it did!

I reached over and tried to make sure the pumps were both on by pulling on what was left of the handle the best I could. On the run, the car had unbelievable tire shake, at about 1000 feet the barrel valve in the fuel system broke right off of the engine and right at the finish line the steering arm broke! Luckily, we were running the small-diameter front wheels, so I was able to keep the car out of trouble and got it stopped safely.

What was really amazing was that we were only allowed to bring a small toolbox—not the big toolboxes we normally brought to the races—with us because of the space limitations we were under. Rahn fished through that little toolbox we had with us, and wouldn't you know there was another handle in there!

So the guys were working on the car to get it ready for the next run, and as it started to get dark, the Japanese were on the track racing their own cars. There was a Japanese Funny Car out on the track, and suddenly the crew got into an argument with the starter. Before you knew it, there was a big bunch of Japanese racers at the starting line, and they were threatening to beat the daylights out of the starter! That was really strange!

By then it was about 5:20 p.m., and it was pretty dark. The track at Mt. Fuji had no lights for racing at night, but we were told to go up and get ready to make our run. We towed to the starting line and the argument was still going on. Rahn went right up to them, told them to break it up, and in no time they were dispersed.

I finally got to make my run with no lights on the track, and in all this darkness the car ran 5.85 at 285 miles an hour. It was the quickest and fastest pass ever in Japan, and that record still stands as of this moment.

At the end of the racetrack, I stopped the car, and as I looked down the return road, here came Melvyn Record on his mini-bike. He pulled up, got off, and went absolutely nuts!

"That is the greatest run I've ever seen in my life!" he says as he was literally rolling around on the ground.

That run pretty much bailed the whole race out, because up until then, the other NHRA racers hadn't really done much.

At Mt. Fuji, Japan, I set the e.t. and Top Speed records for that country—marks that still stand today. *Shirley Muldowney Collection*

The man who predicted our perfect run a week earlier was in Mt. Fuji that night, and after that pass, he came down from the tower and said to us, "You made the perfect run."

When we got onto the bus that brought us all to and from the track at the end of the night, all the racers—including Craig Treble— stood up and gave us a standing ovation, and I've always been a fan of his ever since.

Let me just say that Byron Hines was there, who has been racing motorcycles for a long time, and he came up to me and said, "Shirley, you've got the biggest balls in Japan!"

He was impressed, and that really impressed me.

The Pursuit of Sponsorship

Trying to get a sponsor is an unforgiving process. It's something that can turn racers against each other, and I've seen it happen. There are some racers who will do anything—and I mean anything!—to get a sponsor, even if it's a sponsor that another racer already has. It's just the nature of the beast. But I think that comes with the territory in any form of motorsports.

There are racers who just *have* to be out there. I'd call them desperate. They say, "Hey, let's get five guys together, throw a car together, and go racing because we've just got to be out there!" They do whatever they need to do to get to a race like Indy so they can be a part of the sport's biggest event. That's not the way to race.

There have been a lot of "one-hit wonders" in this sport. Racers who are here today and gone tomorrow.

I read about a racer from the West Coast—I won't mention his name because I think he's still hurt from a racing accident he suffered—and he said, "I'll go broke if I have to for me to go racing."

Oh, God. That's just the kind of mentality we need. "I'll sell the farm to go racing." It's nuts. And he really meant it. That's how people like that feel. It's definitely not the way to approach this sport.

I'll tell you another thing. There are many racers—some of them very good racers—who have given their act away. You don't think that corporate people talk about this out on the golf course?

"Hey, I put a deal together with so-and-so for $100,000."

"Really? I got him for $50,000!"

These corporate people talk, and the racers who give themselves away are ruining it for everyone else. Don Garlits was one of the biggest offenders. He'd give his act away. He'd put too many sponsors on his car for little or nothing. Then, sponsors figured that if they could get Garlits for that little, everyone else had to be worth less.

I'll tell you about a deal that I could have had that I turned down.

In 1977, Marvin Rifchin of M&H Tires in Massachusetts came to me and offered me $25,000 to run his tires exclusively on my Top Fuel car. I had run M&H tires on my gas dragster, and I may have raced my Funny Car on them—I don't remember—but $25,000 was quite a bit of money in 1977. But I turned him down.

At the time, I was running Goodyear tires. I didn't have a financial deal with them, but I think they were supplying me with tires, and I knew I couldn't win with the M&H Tires. Leo Mehl, who was the head man at Goodyear Racing found out that I had turned Marvin's deal down, and he never, ever forgot it. And I was with Goodyear until the cows came home.

Leo was just blown away that I stayed with Goodyear just out of loyalty to his product. Everyone had their hands out back then, and that wasn't what kept me with Goodyear. I went on to win the World Championship that year, and Goodyear played a big part.

My deal with Goodyear ended on November 9, 2003—the day I retired from driving.

The Law Won

We had just finished up a match race at Atco Dragway in New Jersey a few years ago, and we were out on the turnpike in our tractor-trailer. There was a fair amount of traffic, and there were lots of big rigs on the highway that day. Suddenly, this guy whipped by us in his car and he had to be doing 120 miles an hour. A moment later a police car with its lights flashing raced by, and then another one, and then another one. Five, 10, 15—I mean to tell you there was police car after police car going past and they were flying!

Rahn said, "Well, there's a tollbooth up here soon. This should be interesting."

So we kept going, and we were anxious to see what might have happened up ahead at the tollbooth. When we were approaching the toll plaza, all we could see were flashing lights. They were everywhere! There had to have been a minimum of 25 police cruisers all stopped at the tollbooth. I guess the cops can overdo it sometimes.

It was quite a scene when we got up to the tollbooth. They had already gotten this guy out of his car, and there were cops everywhere, standing around, gabbing, and just watching what was going on with this guy they had just grabbed. Once they got him out of the car, his pants were falling off, and it was good for a chuckle.

As we were pulling by, I rolled down the window and called to this guy, "You stupid son of a bitch! Whatever made you think you could get away from all *this*?!"

The cops turned around and saw my name on the trailer, and when they heard what I said, they were practically on the ground from laughing so hard.

Autofest 2000

On New Year's Eve 1999, optimistic promoters organized a multifaceted automotive expo at Moroso Motorsports Park in West Palm Beach, Florida, dubbed Autofest 2000. The event was to feature drag racing, car shows, and a host of other activities and attractions for the tens of thousands of spectators expected to attend. The highlight of the event was a special match race between Shirley and "Big Daddy" Don Garlits that would be held just as the clock was striking midnight, thus bridging the old millennium and the new one. Because of questionable planning, overextension of capital, and the Y2K scare—which affected the attendance—the expo was a financial disaster. Shirley recalls her own memories of Autofest 2000.

Auto-*Theft* is what it should have been called. The fans didn't support it the way everyone thought they would, and it didn't help that a bunch of shysters were involved in it. It was a nightmare for a lot of racers who are still waiting for their money.

The promoters were spending money like drunken sailors, and they would have needed a huge turnout of fans

At Autofest 2000, I raced Don Garlits into the millennium and won. *Steve Gruenwald*

to just break even. When they called me to come down, I was smart enough to get a little more than half of my money before we even pulled out of my driveway. Some racers wound up with nothing.

The Millennium Drag Race itself was actually pretty exciting. There was a lot of anticipation and a lot of drama. *Tense* is the word that I think describes it best. We pulled up to the starting line as midnight was approaching. The fans were all revved up, and the atmosphere was electric. The race went off just as it had been planned with the tree going green a couple of seconds before midnight, and the two of us crossing the finish line a couple of seconds after midnight.

Just before that run, John Force announced he was going to pay $10,000 to the winner. The irony of that was Force was one of the racers who didn't get his money that night. They hooked him for $50,000!

But I won the race against "Big Daddy," and afterward, he was very nice and came over to drink some champagne with us. But later on he was quoted as saying he thought the $10,000 was to be split between the two of us! And he didn't know about the $10,000 bonus until after the race! But that's "Big Daddy"—always on the lookout for his piece of the action.

It was a shame that so many people lost money on that event. The promoters filed for bankruptcy, and some racers will be waiting for the courts to settle the lawsuits that resulted from the financial disaster.

I think everyone who had anything to do with that event will have some unhappy memories for a long time.

Heat of the Moment

Race car drivers are human, and there are times when things are said in the heat of the moment that wouldn't be said in other situations. NASCAR drivers have an advantage in that after the race, they have a cool-off lap or two to gather their thoughts before the cameras are staring at them. We don't have that in drag racing. The fans are always right there, the cameras are always right there as soon as you get out of the car, and you don't always say the right thing. I would never, in my wildest dreams, say that I'm the best interview in the world—and I know that—but I'm never going to lie to you.

I don't know any other way, and maybe you can say I don't know any better.

Paul Page interviews me in 1982. *Long Photography Inc.*

No Personality Allowed

Things are so much different now than they were back in the 1970s and 1980s when it comes to how the drivers are allowed to be themselves. They're not allowed to be personalities anymore. They get fined and they have points taken away. In drag racing and NASCAR, it's the same. They took points away from Dale Earnhardt Jr. not long ago for using a rather tame word that people say all the time. He had just won a race and was really excited.

The corporate world wants a certain image. It makes for boring interviews and stories that are far from factual and ties drivers' hands behind their backs. I think drivers should be allowed to express themselves honestly, instead of having to worry about getting fined for telling it like it is.

The Anonymous Idiot

I was at an AHRA race—I think in Kansas City one year—and this idiot—I don't remember what his name was—was driving a Funny Car. The car got away from him, and it was pointed right at the guardrail. It hit the guardrail, and he climbed out of the car while it was still running!

The car was butted right up into the guardrail, and the rear wheels were turning and smoking as he was climbing out of the side window! That's what I mean when I always say, "It's a wonder more people didn't get killed." It is amazing to me that the good drivers—the ones who really knew what they were doing—survived those years and didn't get killed by the ones who didn't.

Ask Don Prudhomme. We've talked about it every now and then. He never got hurt in all the years he was racing in Funny Car and Top Fuel. In fact, at the fall NHRA Chicago race in 2004, I was walking past his hospitality area next to his Top Fuel and Funny Car teams and saw him sitting up on a table down at the far end. I figured I'd go over and join him, so we cracked open a bottle of red wine, we both had a cigar, and we talked about the old days!

He said, "You dodged a bullet, girl."

If anyone knows about dodging bullets, it's Prudhomme. Like I've said: I saw him do some amazing things when he was a driver.

And nobody had a stronger desire to win than he did. It didn't matter who was in the other lane or where he was racing. He's still that way and he deserves a lot of respect.

Dragsters Versus Funny Cars

There are some people who believe Funny Cars are safer than Top Fuel dragsters. In a Funny Car, you may be safer, but I'm not really sure how. In a dragster, you've got that long front end in front of the cockpit that can break off in an accident. That reduces the weight, and so it also reduces the energy in an impact. In a dragster, you're sitting low, and you're more protected by the rollcage around you.

In a Funny Car, you're sitting up higher inside of this confining cocoon. It may seem like you're protected in there, but let me tell you, that's a false sense of security and always has been. Funny Cars are prone to have fires, and you can't see very well in there, especially at night. But the drivers out there today who drive those Funny Cars are really, really good. The drivers who were in those cars back then were not that good.

You had your "Jungle Jim" Libermans and John Forces—of course, if you go back far enough, I can remember when Bill Doner, the great West Coast drag racing promoter, would pay Force *not* to go down the racetrack! Force will still admit that to you. Nobody else would, but Force would.

But in my opinion, a dragster offers you a little more protection. However, Funny Cars are a hell of a ride! I can't imagine what it's like to go 4.66 at more than 330 miles an hour in one! In a way I can, but then again, I can't. I would have jumped into a Funny Car in a New York second in later years because I know they're fun to drive.

Both classes have gotten better overall when it comes to safety, because the people who are working on them have gotten a lot better than those who were working on them years ago. Fuel cars are only as safe as the people who work on them.

I'll admit that the wing in a Top Fuel car could be its weak link, but when Jimmy Nix was killed, it emphasized the risk in going Top Fuel racing without a fulltime crew. That wasn't a car that was taken apart and checked out after every race. It was run, then it went on the trailer, and it was towed so it bounced and bounced and bounced some more as it went down the road. How many people who worked on that car *lived with that car?* That car was never taken apart every week after it was run like Rahn; Lee Beard; and people like that do, looking for problems before they look for you.

That didn't happen with Jimmy Nix's car, and it bit him.

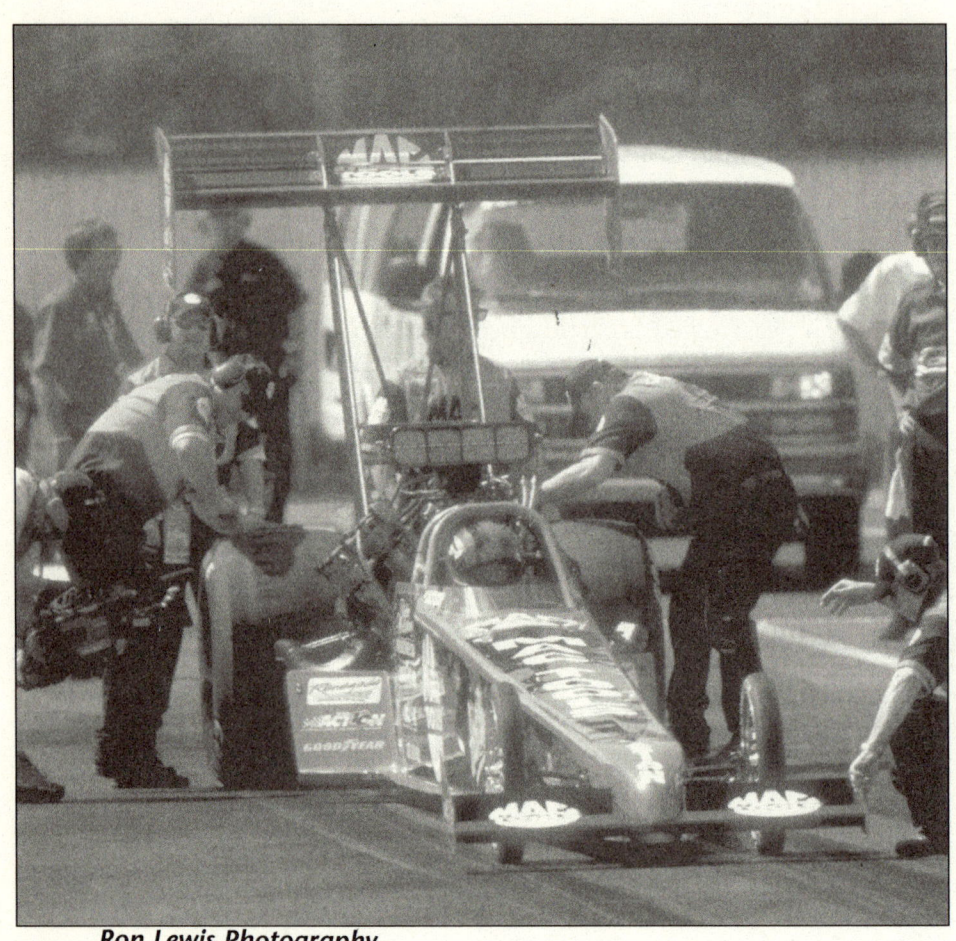
Ron Lewis Photography

Chapter 3

The Characters on the Circuit

"Hucksters"

I was pretty well liked at Fonda when I started racing up there in the beginning. The fans loved me. The owner of the track in South Glens Falls where I raced a lot in the early days was a big fat guy who, I guess, fancied himself as a driver, but everyone called him "Fat Jack." I wasn't sure if he was for me or against me, but the owner of New England Dragway in Epping, New Hampshire, really was a bastard. "Sy" Sidebotham was his name, and I can remember racing up there in the 1970s and having to wait forever to be paid.

The race would be over, and we'd all be standing outside the tower, and he'd be inside counting the money. That's all he cared about, and when someone told them we were all outside waiting for our money, he'd say, "Let 'em wait." I mean to tell you, I'd be there, Don Prudhomme, Connie Kalitta, Jim Liberman—all these racers would be waiting for their money, and it was almost as if he wanted us to beg for it.

Jack Doyle was just as bad, and Gil Corrain was another one up there. We called them "the partners in crime," and we meant it. In later years, I often wondered how the sport ever survived with people like that in charge. They were such hucksters, such "carnies," and I thought that element had been long gone. Bill Bader of the IHRA who owns Norwalk Dragway in Ohio is like that. It was never about racing; it was all entertainment; it was about causing a big scene. I'll have more to say about him later.

Whenever I see these scantily clad chicks backing up the race cars today, the first thing that goes through my mind is that went out in the 1970s. That was yesterday, and we are what we are today. The sport can stand on its own. We don't need that anymore.

In the 1970s—hey, I wore hotpants, too. Today it's not about the sideshows. It's about the drivers, the cars, and the performance.

Wild Blue Yonder

In 1972, Connie Kalitta had a twin-engined Cessna airplane. He then stepped up and bought a twin Beech. That plane wasn't a "tail-dragger"; it was a tri-wheeler freighter. I spent quite a bit of time flying with him back then. I remember when he first got it, he had a pilot flying with him who turned out to be a real idiot and landed it at City Airport in Michigan—without the landing gear! The crash wiped out both engines. Connie was wild!

I was back home visiting my mother in New York when he called me and told me what happened, so I drove the 500 miles in eight hours back to Michigan. Three days before Christmas, Connie had me take his pickup truck with a twin Beech airplane engine hanging out of the back of it and drive in an ice storm to Illinois to have the engine repaired. Somehow I made it OK, sat in a motel for three days while it was being worked on, and eventually got the engine back to him.

I was there for Connie in those days and had a couple of hundred hours with him in the air, flying in the left seat, many times while he was asleep in the back—with no autopilot! He taught me how to read the maps, how to read the instruments, and how to keep the altimeter within 100 feet of the proper altitude. I took pride in being able to do that.

I was told that I should wake him up if it looked as if we were getting at all close to other aircraft or something out of the ordinary happened. I'd rock the plane and flash the lights to wake him up. That was the signal to him to come up to the cockpit.

One time he came up and sat in the right seat.

"OK, I'm going to shut an engine down and see if you can hang onto this thing," he said.

So he shut an engine down to see if I could maintain control of this big airplane until he had enough time to get to the front. Well, I did and I had no problem.

"Are you satisfied with that?" I asked him.

"Yup."

And he went back to the rear of the plane and closed his eyes.

I never had a problem in all the times I flew the plane for him. He trusted me.

"Broadway" Bob

Bob Metzler is another huckster. He ran Union Grove Raceway up in Wisconsin. When I hear all these accolades for him and how he's being inducted into the Hall of Fame, I want to be ill. This is a guy who divvied out checks that weren't any good. That's not to say that he never made good on them, but I had to chase him more than once because the checks he paid me with weren't good. I had to find him, and sometimes when I did, he was drunk as a skunk. It is still amazing to me that drag racing survived in places where he ran the show.

Some of his events were deathtraps. The fans weren't safe, and one of the most dangerous things we ever did in drag racing was run up there.

"Broadway" Bob made a ton of money. I'd come into Union Grove early when I raced up there and he'd run me around to the radio stations and newspapers on a schedule that was endless. I made a lot of friends in the media then because it was always easy to get interviews when I came to town, and I treated them as good as they treated me.

But it was a long day when I made the rounds for Metzler. We'd leave at 8 a.m., and do you know what lunch was? He had the nerve to stick a quarter in a vending machine and hand me a candy bar! That was lunch.

So I'm not a big fan of "Broadway" Bob. I still run into him at various functions, and I basically tolerate him. I don't go out of my way. He came to me a while back and asked me if he could link on to my website.

I said, "If you value your life, you'd better not."

"Moon" Mullins

In the early days, sometime in 1973, I got lucky enough to get in touch with a man whose name was Bernard "Moon" Mullins. He was a wonderful, wonderful man. He

Shirley Muldowney Collection

was the public relations man at Chrysler, and he knew that he could get a whole lot more return on investment with me than with anybody else in the sport at that time. Chrysler was much more interested in racing street-type cars and never really went full bore with nitro cars except with me and Roland Leong and the *Hawaiian*, because they could see what it could turn into.

Even today, when you see archived pictures, which are used in magazine features on my career, you can see the Plymouth logo on my dragsters. They're still reaping the benefits, but back then they nickel-and-dimed me, they tied "Moon" Mullins's hands, and he went out on a limb more than once to help other drag racers and me. I remember he helped others, including Roland, by getting him engine blocks and bodies.

Finally, after two years as Connie Kalitta and I were driving this old Ford truck that he had bought second-hand, which we used for towing—mainly because our trailers were getting bigger and heavier and my Chevy truck wouldn't pull them anymore—Chrysler told us they were going to give us a truck, and they did. They gave me a Dodge club cab, not a crew cab. It didn't have a radio, and it didn't have air conditioning.

We used that truck for a couple of years, but it was always a pull and a tug to get anything from Chrysler. Meanwhile, we always showed up in their press kits and they'd toot their horns whenever they could when they sent me to auto shows in Detroit or anywhere they felt I'd be some benefit to them. But for me, there was no "R.O.I." or return on investment. I found them to be cheap whenever it came to taking care of me. But when

I began a marketing relationship with Chrysler in 1973, and the ads from the era still show the Plymouth logo on my dragsters. *Chrysler Historic Archives*

they unceremoniously let "Moon" Mullins go while he was sitting in a golf cart at a golfing event that he had helped promote, that was an insult.

The Good Outweighed the Bad ... Usually

Keep in mind that I worked for many wonderful people over the years. There are only a few who didn't pay me despite a great deal of effort on my part to get my money. Very, very few.

I have no use for motorsports promoters who are not upfront and honest with me. They get one shot, and that's it. That's the way I am. If you sting me once, it's over. It wasn't about the money as much as it was about my self-respect. That was at the top of the list. You need money to survive, sure, but how I was treated had a lot to do with whether I would do business with someone. I'm still that way today.

Don Prudhomme

I have always been a big fan of the way Don Prudhomme raced. He always raced to win, and just like me, he never got over the feeling of disappointment from losing. Don is a survivor and has had to make many decisions that have allowed him to reach the levels of success others have never attained. Sometimes it meant having to play the game, but the way that I look at it is simple. Whatever

Don Prudhomme and I share a moment before a race in 1977. *National Hot Rod Association/National Dragster*

games he played all those years, he backed it up on the racetrack. Bar none. You could knock all the other stuff he did, try to pick it apart, and make of it what you want, but you can never take away all of the great things he did as a racer.

Back at the World Finals one year when they were held at Orange County Dragway in the 1970s, I saw him get to the finish line in his Funny Car just as the engine exploded. It was a hell of a fire. I had never seen anything burn like that in my life. But he got that car to swerve—just slid it sideways—and it came to a stop as quickly as I had ever seen anyone do it. He jumped out of that car in no time, and it was the greatest driving job I'd ever seen. That's the kind of skill he had as a racer.

And I can remember how he would make low rpm burnouts, which nobody else could do. He just had a level of finesse when he drove that was far above that of a lot of other guys who considered themselves great drivers.

Jon Asher

Jon is a great drag racing journalist and photographer, whom I met at New York National Speedway in 1970. I was running my twin-engined gas dragster.

He said to me, "Man, you've got to build a Top Fuel dragster!"

I probably should have taken his advice because I soon wound up in a used Funny Car.

Asher was always there. I would ask him for a picture from time to time or ask him to write me a bio that I needed, and he would never say no and never send me a bill. He wrote proposal after proposal for me when we needed sponsors and made personal appearances with potential sponsors and did a hell of a job. He's very well versed and a great journalist and someone who will take the powers that be to task when they have it coming. He's oh so tough, but oh so smooth.

Asher and Al Eckstrand gave me the same advice. They both said, "When you're mad, don't put anything in writing."

So whenever I was upset and needed a letter written, I'd either call Asher and say, "Jon, I need a letter," or I'd write one and have him look at it. A lot of times when I did that, he'd look at it and say, "Well, we really need to make a change here or there." He was able to help me get the point across diplomatically. I'd want to let someone have it right between the eyes, and more than once, he saved me. Asher had that ability, and I've always been grateful to him for that.

Sometimes after something happened and you moved down the road, you'd look back and realize it could have been handled a bit better. He understands that, and he's been there for me in those situations more than once. He helped me calm down.

Rain Money

There were a few times that track owners or promoters wouldn't want to pay you for whatever reason.

I never ran at York, Pennsylvania, but we went there in 1978 after they booked us for a match race. We showed up there after there had been a horrific downpour, and on the way to the track, we had to pull over next to the road because it was raining so hard and we couldn't see where we were going. Obviously the race was rained out.

We finally arrived there, and all the racers were pulling things together and loading everything up on their trailers. There wasn't going to be a race that day. So, I went up to the tower and walked up to the third floor to collect my rain money because we had driven all the way from Detroit.

The promoter at the time was sitting at a table up there, and there were several other fellows whom I didn't know sitting around with him. There were several big long card tables set up in the area, and there had to have been about 50 Winston ashtrays stacked up on those tables in stacks of five. I asked the promoter for my rain money.

He said, "Get outta here! We're not giving you anything!"

Let's face it, they had probably taken a hit financially with the rain out, but I wanted my rain money regardless. Well, we started shouting back and forth, and when it became obvious that I wasn't going to walk out of there with my $500, I picked up those ashtrays and started sailing them! And those guys started ducking!

They were running and ducking from one desk and table to another, and I cleaned every ashtray off that table. When I walked out, I saw a telephone hanging on the wall near the doorway, and I pulled it right off the wall. But when I did, the handset sprang back and hit me in the head and gave me a little cut, which made a little blood trickle down my face. When I got downstairs, Rahn was coming in, and when he saw the blood on my head, he wanted to go up there and clean the place out! I had to say, "No, no, it's OK!"

But I was going to get my $500 one way or the other, and I did.

Later on Connie Kalitta had heard what I had done and said to me. "Here's some advice. Don't ever close a door."

And from that point on, I always tried to heed his advice.

Gary Beck

Gary Beck won the NHRA Top Fuel World Championship in 1983. He was always nice to me, and the story I remember most vividly about Gary was at the 1980 World Finals when there were four drivers who still had a shot at the championship. Gary was one of them.

He had lost in an earlier round, but if I lost in the semifinals, he would still win the championship. He was standing down at the end of the racetrack with his wife, Penny, holding a bottle of champagne. When I won that

Gary Beck was always a class act—even in defeat. In 1975, I beat him at U.S. Nationals, but lost to Don Garlits in the finals. *Shirley Muldowney Collection*

semifinal race and pulled off the track, he walked over to me and handed me the bottle of champagne.

That was a classy thing to do. Like I said, Gary Beck was always nice to me.

An Unhappy Chassis

Ron Attebury built the chassis on my 1997 and 1980 championship Top Fuel cars, but we had problems with him. Both cars were too small for me, and we had to wait forever to get them.

I remember we sold our 1976 car after the last race of the season, which was an IHRA race in Atlanta, and then planned on picking up our new car from Attebury on our way to Seattle. When we got to Fremont, California, the car wasn't finished.

Connie told Attebury, "Finish that friggin' car, or you'll be a dead Attebury!"

Those were his exact words.

We finally got the car out of Attebury's shop and were towing it to Seattle. My son, John, and another member of the crew were working on it in the trailer as we were going down the road.

The 1980 car that Attebury built for me was delivered so late that the paint was still wet when we first rolled it out of the trailer in Pomona. But we wound up winning that race and beat Connie in the final.

I will say that that car was the first Top Fuel car that was really detailed. I'll never forget Garlits coming over while it was parked in the pits and saying, "I've got to take a look at this contraption."

He always had a way with words …..

Jeb Allen

Sometimes I think that people get the impression that I was at war with every racer and every racer had a bone to pick with me. That's not true. I raced against some drivers who were always respectful of me and didn't play the games that other did.

Jeb Allen, who won the NHRA Top Fuel World Championship in 1981, was just a kid back then. But I thought he was a good kid and a good driver. He struggled, too, when it came to having the financial resources to run the team. I'm not sure, but he may have struggled more than I did.

The closest we ever came to having a problem was in Montreal one year when we had both lost and we were back at the Granby Hotel. They had wedding parties booked there on just about every weekend of the year, and the food, the décor, and the atmosphere were really classy. Elegant French wine, elegant French food—the works.

The only problem was that the hotel was seven or eight stories high, and it had a nightclub on the top floor. Those wedding parties made quite a bit of noise on Saturday nights when racers wanted to get some sleep for the race on Sunday.

But the restaurant served excellent food, and on this particular weekend the guys on my team and I had just come from the track after losing, and we were going to have dinner. The restaurant was still fairly empty because it was Sunday and still a bit early. Jeb was also there, and he had had a couple of drinks, and I was sitting at a table with Pat Galvin. Rahn Tobler, my crew chief, still hadn't made it to the restaurant yet.

I don't quite remember what I said, and I can't recall if it was about Jeb, but he obviously thought I had made a comment about him because he marched over to the table where I was sitting.

He said, "What did you say?"

Then he made a gesture—unintentionally—that knocked my glass of wine right into my lap. Pat got up and

Jeb Allen was a good driver who ran into some tough times.
Shirley Muldowney Collection

was about to get into it with Jeb, but I could tell it was an accident.

I told Pat, "It's OK. Sit down. He didn't mean it."

And Pat sat down. That could have been a pretty ugly situation, but it calmed down quickly.

I saw Jeb not long ago, and he looks great. He's a contractor and has done quite well for himself after leaving drag racing. I am still pretty close with his first wife, Cindy. I like her, and we enjoy getting together, but I don't mention it to Jeb on the rare occasions I run into him. I'm not one to say, "Hi, Jeb, and by the way, I had lunch with Cindy last week." I just don't think I need to share that with him.

The Producer

Charles Roven is better known as Chuck Roven, and he's the producer who was responsible for several hit movies. His wife, Dawn Steel, was the president of Paramount Studios and was behind a bunch of other hits such as *Top Gun*, *Flashdance*, and *Cool Runnings*, that movie about the Olympic bobsled team from Jamaica. She was a smart woman. Very strong. I think they hired about 20 secretaries before they found someone who could handle working for her. They called her "Steely Dawn," and that just about said it all.

Word was she had an "automatic door slammer" that she controlled at her desk, so that if you were asked to leave, she slammed the door behind you! Oh, what a tough cookie she was! I thought the world of Dawn Steel! (She died of a brain tumor about five years ago and Chuck was left to raise their gorgeous daughter, Rebecca.)

It was interesting how I met Chuck. Two other people had come to see us with their original idea for my movie while we were racing in Pomona, Herman Zimmerman and his wife, Sandy. He was really a cool guy, and I became very friendly with both of them. I stayed at their house numerous times while I was in Los Angeles, and they treated me like a queen.

He was a set decorator at Paramount, and so I gave them the option for a year to try to put the deal together to have the movie picked up by a studio for production. But they just couldn't sell the movie—they just didn't know how. The week before the World Finals in 1980—the deadline set for them to sell the film pitch—the Zimmermans really tried everything to get the movie sold. They tried to have a TV movie made. Anything that was a possibility, they pursued. But they just didn't have any luck. But they were sympathetic, because at that time, I didn't have a lot of help in finding a sponsor, so they introduced me to Chuck Roven.

Chuck Roven came to the finals along with a cohort of his, Claude Revier, a gentleman with an international bearing about him. (Claude drove the sleekest Ferrari you could buy, and it had Arab license plates. In fact, I was responsible for him getting together with Kathy Lee Crosby when I was approached to take part in a *Battle of the Sexes* television show. He and Kathy Lee saw each other for a while and were a cute couple.)

Now Chuck was a real professional, and I'll tell you why. Up until that weekend of the finals, he had waited until the Zimmermans had made all of the efforts they could to get my picture sold in the time frame we had agreed to. But the weekend of the race, Chuck was free to represent my movie, and he did. He was both a movie producer and a manager. He was well to do and had connections in a lot of places.

Two days later, I was having dinner in Beverly Hills with Chuck and the people who bought the movie, Aurora Productions. They handed me a check that night for $25,000 to show me they were serious.

The thing I remember most is what he said to me at the time, and I've followed his advice ever since.

He said, "I don't care what you do, but you must promise me that you won't sign anything—*anything*—until I see it first!"

He has a history in Hollywood that goes something like this. He will change the word "this" to "them" on page 62 of a 65-page contract. Talk about a "detail guy"!

But the NHRA hated him.

I know that there were people there at the time who attacked him all because they were afraid to deal with Chuck Roven. He was simply too big for them. They didn't know how to deal with him. So I took a lot of flak from the NHRA because of Chuck Roven, but he was a friend and he'll always be my friend. He's a good guy to know.

In the end Chuck Roven made sure I received every single penny we contracted for. *Heart Like a Wheel* did more to boost my fame than practically anything else in my career—with the exception of Rahn joining my team

Heart Like a Wheel gave my fans a dramatic inside view of my life. Bonnie Bedelia (right) played me in the movie. *Shirley Muldowney Collection/Long Photography Inc.*

and giving me my most successful years on the dragstrip. You could never replace that. And can you tell me anyone else who came along and made the most of that opportunity? Chuck helped make it happen.

I have absolutely no horror stories about my experiences in Hollywood. I met some really terrific people.

Crash of the Stuntman

I have a quick story about Hal Needham, the stuntman who got pretty big working with Burt Reynolds on those *Smokey and the Bandit* movies.

In 1982, Needham and that crazy comedian, Jonathan Winters, crashed the sports press conference in Los Angeles where Steve Earwood of the NHRA and I were promoting the upcoming Winternationals in Pomona. Needham just charged in and basically took over the stage. He and Winters had some cockamamie schtick going on, and you could tell that Needham was stewing. He wouldn't even speak to any of the Hollywood people who were there that day. He was rude, he was curt, and he was pissed because he hadn't been contacted to have a part in *Heart Like a Wheel*.

He thought he was "Mr. Motorized Anything," and he just about forced us off the stage because he wouldn't acknowledge anything we were going to do. I'll never forget what a gentleman he was not.

Self-Defense

I'll be the first to admit I wasn't always the most well-behaved racer out there. But I want to say flatly that I never cold-cocked anyone, and I never attacked a fan. That's not to say that my people didn't have to take a few unruly folks down. Many times over the years we had to defend ourselves for whatever reason and didn't hesitate to do it.

I detested the tracks that served liquor, especially the old IHRA "beer tracks." I hated them. Drunks always found their way over to where we were.

I was so glad to have my old friend Donnie Couch with me for many of those years. Pat Galvin, too. And my crew chief, Rahn Tobler, was never one to allow things to get too far out of hand. Charlie Lendrum was another friend of ours who was part of the team for a while. Charlie was a trip.

He'd drive the truck for us, and we'd all be riding along—John, his friend Brian Davis from Michigan, Rahn, and I. Rahn would be stretched out in the sleeper and suddenly poke his head out and ask, "Charlie, how goddamn fast are you going?!"

"Eighty-five!" he'd say.

And Rahn would say, "Slow this mother down!"

But that's how he drove. Foot to the wood and he'd live on coffee and cigarettes. That was Charlie.

A Not-So-Quiet Dinner

Back in 1982, winning Top Fuel at an NHRA national event paid $7,500. They also arranged our hotel accommodations, and in Brainerd, Minnesota, they had us staying at this resort that was way downwind of the racetrack. I mean, it was out in the proverbial middle of nowhere. We didn't actually have rooms, but rather, this place had us staying in these cabanas, and your allotment of hot water was pathetic. You took a shower by literally standing under a watertank and pulling a cord, which opened up a hole that dropped water on your head! Luckily, we didn't spend much time there because we spent most of our time at the racetrack.

But after we won the race in 1982, we had a celebration dinner back at the big dining room in the resort, which unlike the little cabanas, was really nice. We sat at a large table that was over by the windows that overlooked the lake where the resort was located.

By the time we sat down to eat, it was dark outside, and the dining room was filling up with people. We had quite a gang there, including my son, John; Pat Galvin; Charlie Lendrum; Rahn Tobler; and me. We were there to celebrate our win, and we were looking forward to enjoying our meal and some drinks as we were looking over the water.

Sitting at a table just behind us was a party of six or so, including a fellow and his wife, and this guy was big. We later found out he was an ex-football player who had a drinking problem and had gotten into some trouble around there before. He looked over at us, and when he

I had a very close relationship with the members of my crew. Charlie Lendrum (right) was with me almost from the beginning of my professional racing career. *Shirley Muldowney Collection*

saw we were still wearing our racing uniforms, he started in with, "So, you were probably out at that big race out there today. I heard some girl was lucky enough to win that thing. ...."

We sort of put up with him as long as we could until Rahn or I spoke up and said, "Why don't you spend a buck and come out and see the races? Come out and we'll give you an autograph and you can enjoy what we do out there."

It went back and forth between the tables for a few minutes, and by that time, he had gotten his steak dinner.

The next thing we knew, he turned around holding his steak knife like he was going to attack someone with it and made stabbing gestures, kind of playfully, on our table between John and Pat Galvin.

John saw that and stood right up and backed away from the table. He was scared and moved backward until he was right up against the plate glass window next to the table. Pat reached right over and grabbed this guy's hand until he dropped that steak knife. At that point, we had only been served our salads.

"Calm down, man. ... Stay loose," Pat said to the guy.

Pat stayed cool. But the guy stood up, turned around, reached under our table, and flipped it over! There was food everywhere! Then Pat turned around and flipped *his* table over!

A waitress came out of the kitchen with a tray filled with dinners, and I'll never forget the look of shock she had on her face when she saw what was happening in front of her. Her jaw literally dropped to the floor!

Now this whole time, this guy's wife was just sitting there saying nothing. When their table got flipped over and their food went in every direction, I looked over and there was blue cheese dressing dripping from her nose! She didn't move! She just sat there with this exasperated look on her face. She had obviously been through this before.

And then the melee started.

I stood back waiting to see if I could get a few licks in of my own. Pat had a hold of this guy, but I'm not sure if any punches were thrown. Before the fight broke out,

Charlie had gone over to the bar and gotten a big, green, heavy glass ashtray and had put it on our table. I don't know whether it was because it was so hefty or bulky or what, but that ashtray was the only object left on that table after things got crazy.

I reached down and picked up that ashtray and sailed it as hard as I could right at this guy. and it hit him squarely on the head! You wouldn't have believed how this ashtray shattered into millions of pieces! It was like it exploded!

I was thinking, "Now you can sit down, you son of a bitch!"

And he did. Immediately.

The manager of the dining room, who knew this guy and knew he was trouble when he was drunk, came over and said, "You need to get out of here now. Go now!"

He said he knew who I was, and I knew that meant that the cops were on the way. So we got out of there and headed right back to our cabanas.

When we got there, we began throwing all of our suitcases into the truck and trailer that was parked back where we were staying. We jumped into the red dually and took off for the main road heading out of the resort. It was a two-lane road with one lane in and one lane out. Suddenly, we saw some flashing lights and more flashing lights and even more flashing lights as four police cars flew by us in the other direction. It was deadly silent in the truck as those police cars whizzed past.

I'll never forget, I asked Rahn, "How far is it to the state border?"

And all he said was, "Sit back."

Meanwhile back at the resort, our buddy, Doug Kerhulas, who did not have dinner with us but had gone out to pick up a pizza for himself, returned just in time to see the police looking for the troublemakers. The police, believing he was a party in the disturbance because he was also driving a red dually truck, slapped him into handcuffs and took him into custody. Eventually they let Doug go free, and we all had a good laugh about it later on, because Doug was a pretty big guy and we could have used him when the fight broke out.

(Doug was a real good friend of ours who had been in a serious accident in Columbus two weeks before I had my accident in Montreal. He had a parachute failure and got into the safety net at the end of the track very hard. He sustained a pretty serious head injury, and it gave him some lingering problems, but after his accident we stayed with him for as long as we could while he was in a coma until it became necessary for us to leave.)

A Good Friend to Have

After the scuffle in Brainerd, Minnesota, we made it back to Detroit, and it seemed everything was cool.

About two weeks later, I was served with a summons at my house. The fellow who started all the trouble in the restaurant found out who I was and was attempting to sue me. I picked up the phone and called Charles Roven.

I told him about the problem we had in Brainerd and how my team and I were just victims of circumstance. He listened to what I had to say and told me to stay at home and he'd have somebody call me. Charles's secretary called me soon and said that I would be hearing from someone who would take care of it for me. So I waited.

It wasn't long before I got a phone call from someone who had been a very good friend of the Roven family for years and who was a competent attorney. I had met him, in fact, when *Heart Like a Wheel* opened in Beverly Hills. He asked me the details of what happened in Brainerd, and I gave him a complete account of it. He got the information from me he needed, made three phone calls, and that was the end of the deal. I never heard another word about it.

The attorney was Robert Shapiro, who you'll remember was part of O.J. Simpson's original defense team. And are you ready for this? When I received his bill, it was for $7,500—the exact amount I had won in Brainerd before we headed to that restaurant to celebrate!

Johnny Carson

Susie Arnold, who is now the PR person for Kenny Bernstein but was with the NHRA years ago, took great pride in saying she set up my appearance on *The Tonight Show* with Johnny Carson, but she really wasn't the one responsible. That's not how it happened.

She was at the NHRA in 1986 at the time the people at *The Tonight Show* called and wanted to get in touch with me. They didn't come through a publicist or anyone else.

They wanted to call me and invite me to come on to the show. Man, I was nervous! You couldn't imagine how nervous I was!

I didn't know what they wanted me to wear, so I went out and bought myself a $300 sweater to wear on the show. I wanted to wear that sweater so badly! But they told me they wanted me to wear my firesuit.

I said to Freddie De Cordova, Johnny's producer, "You know, all my career everybody has asked me to wear that firesuit whenever I make a TV appearance, but I really would like to wear the new sweater I bought."

Besides, they didn't understand that a firesuit has several layers of Nomex and under those TV lights, I would have been cooked alive!

Freddie was such a great guy, and he said, "You can wear whatever you want."

They didn't argue with me, but later on I could see why they wanted me to wear the firesuit.

When I got to the studios in Burbank, California, I was taken to the Green Room, and my name was on the door. It was my own special room to wait in until I was to go on, and there was a lavish spread of food on this table. Refreshments and hors d'oeuvres and anything you wanted. It was really wonderful. They didn't overlook anything.

One thing that Johnny never ever did during all those years he was hosting the show was meet a guest beforehand. He wouldn't talk to them or see them until they were actually introduced on the show and joined him on the set. When I was in makeup, Johnny came down to meet me. That was probably a one-time thing for him. I knew he

I got to go on *The Tonight Show* with Johnny Carson in 1986. *Shirley Muldowney Collection*

had already been in makeup himself because I saw these little white lines that had been applied under his eyes, and I was told he was very much into having them because they added brightness to his eyes when he was on camera.

I had the chance to go out while the studio was still empty and take a look around and then went back to the Green Room. At the time, I was wearing high-top shoes that laced up because they gave me a lot of support. This was in 1986 when I was still recovering from my accident in Montreal and I could barely walk, even with a cane. I was sampling all the food they had set out for me and socializing with some of the people who came to say hello to me in the Green Room when, suddenly, Freddie De Cordova comes in and says, "You're on!"

But my high-top shoes were all untied and I said, "I can't go on now! My shoes are untied!"

Well, Freddie got down on his knees right there in the Green Room and helped me tie my shoes! I walked out on stage and it was wonderful!

Oprah Winfrey was on the show the same night that I was, and she didn't get half the time with Johnny that I did. But the big attraction was the golf cart race they had set up between Johnny and me backstage. I was told that Johnny had practiced all day long getting ready for this race. He just wasn't going to allow me to beat him!

They had a backdrop behind where we were going to race that was painted like a crowd at a drag race with fans cheering and jumping up and down. It was funny!

Something else that really cracked everyone up was that Johnny was wearing a pair of racer's overalls with about 900 sponsor stickers on them, which explains why they wanted me to wear my firesuit.

Anyhow, Johnny beat me. He got me off the line and won. People asked me for years afterward if I let that happen. Of course not. I didn't let him beat me. Afterward, they cracked open the champagne on camera, and it was a lot of fun. I told Johnny I'd love to personally invite him to come to a race and be my guest in my pits. I told him I'd let him sit in the car, warm it up, and feel what it was like to be sitting in front of 7,000 horsepower. His eyebrows shot up when I said "7,000 horsepower," and he had the audience on the floor! Really, it was one of the best times I've ever had.

Keeping Sponsors Happy

I never had a problem with my sponsors. They knew they were getting a big bang for their buck. "Moon" Mullins, the PR director at Chrysler, told me that I was the best deal that Chrysler ever made back then.

Performance Automotive Wholesale was a pretty good deal for us. The owner, Keith Harvie, was an Englishman and a very good businessman. He had a somewhat strange kind of charisma, but he was with us for four years and his money was always there. It was just hard to figure out his personality. One time, he really hurt our feelings.

One time, we got back from an event, and it had not been a good outing. We may have failed to qualify, I'm not quite sure what happened, and when we got home, we got a message on our answering machine from him that said in a very off-handed way, "Hi, this is Keith. What do you think? Garlits turned a great e.t. and Top Speed up at Bakersfield this weekend!" *Click.*

He did the wrong thing when he did that. I was glad when 1989 ended, and the deal was over. But once again, his money was always there when we needed it, and we had no complaints with that part of the deal.

English Leather was more trouble than they were worth, because they were heavily into the promotional appearances, displays, videos, and all those things that ran

P.A.W. was a very good sponsor for us. In this car I became the first female driver to race in the 4's. *Auto Imagery Inc.*

us ragged. The owner of the company, Gay Mayer, was a very nice man and very stylish, but his PR people put us through the wringer. They asked us to bring the race car into New York for a promotional display. Try bringing a transporter into Manhattan in all that traffic and see what that's like!

English Leather had an in-house promotions director named Harvey Chandler and there was not a bigger knothead out there! He finally invited us to a Manhattan lunch one time, and when he pulled up to the restaurant in his car, the parking attendant said to him, "Hey, you've got a dent in your car. I can fix it for you."

Harvey said, "Can you fix it while we're having lunch?"

The attendant said, "Yeah, sure!"

And this guy, Harvey, believed him! While we were having lunch, the guy bondoed that fender from one end to the other! It was unbelieveable! And the paint was about 15 shades darker than the rest of the paint on the car. It was a mess!

Harvey paid the guy, and none of us ever said a word about it. He was a big PR hotshot, who was making it in the automotive world. He was trying to impress us. Oh, he impressed us all right!

Al Eckstrand

Al Eckstrand was a great Super Stock racer driving for Chrysler, who was a good friend of mine in the early days. All of his cars bore the nickname *Lawman* because he was an attorney by trade, specializing in probate law.

He's 75 years old now and spent 25 years in England restoring old castles. He's had such a fascinating life. I hear he's been honored by the Queen of England, he's known within the royal family, and he's such a worldly, brilliant man. He travels around with one of his *Lawman* racecars giving presentations for the armed forces. He's a dear friend and a heck of a guy.

He always tried to help me in any way he could. He never represented me, but I'll never forget the advice he gave me when it came to legal matters. Always say, "I think..." or "I feel..." and you'll be covered legally. He always advised me, and even today, all I have to do is call and he's there for me.

The Meanest of the Mean

I think the guy who could be the meanest—but never, ever to my face—was Richard Tharp.

He had a cunningness and a tricky, calculating way about him that said, "I'm gonna trick ya!"

I once thought that cunning had a positive connotation to it, but it's really not a word I like to use to describe someone.

I'd have to say Lee Beard was like that. He was the kind of person who I think would waste his time trying to come up with ways to get in my head and give me a jab. The story I always think about is what happened when we were returning to the NHRA in 1997 after we had been away since 1990.

In 1991, we had a really disastrous episode with a company called Petromoly, an oil company based in Houston. We had a $4 million deal—signed, sealed, and delivered—except the owner of Petromoly never delivered. He made the deal with me because he was planning to sell the company to a group of Venezuelans, and I was the frosting on the cake. If he had a sponsorship deal with

Richard Tharp was the trickiest racer on the strip. *Auto Imagery Inc.*

Shirley Muldowney, it made his company look a little more attractive.

Well, we ran two races and received a total of $3,500 from him! There was a clause in our contract that stated if there was ever a disagreement between us, it would have to go to arbitration. I had no choice but to go after him. We had a deal, and he was trying to back out of it. That arbitration cost me $25,000.

After that horror show, we eventually made our return to the NHRA in Dallas in 1997. When we got there, an airplane flew over the racetrack towing a banner that said, "Hi, Cha-Cha. How's the Petromoly deal?" I never would have known who was responsible but Kim Lahaie-Richards came over and said Beard and Dave Settles were the ones who hired the plane.

Rahn was furious! He went over to Kenny Bernstein, who Beard was working for at the time, and told Kenny, "This is what your crew chief did!"

Kenny wasn't completely convinced and said, "Well, I really don't think Lee is responsible."

Rahn told him, "I know so."

When we got into the staging lanes, I was sitting in the car and watched as Pat walked over to Beard. Lee was shaking. He looked shook up. I give credit to Pat for not putting his hands on Beard, but he did say something to the effect of, "Hey, these people are working hard with their backs to the wall trying to come back and run out here. Don't you have anything better to do with your time?"

I was waiting for the NHRA to clean up an oil-down while all this was going on, and when Pat had finished talking to Beard, Lee walked past my car. He was just a few feet away from me.

I called, "Beard!"

He turned around.

When he saw me, I said, "You've got a lot of god-damned nerve!"

I just laid it on him. And he turned around and just kept walking.

What was it that made Beard not like me in the first place? I really don't know. It may have been my success. Remember, to this day he's only won a single championship and that was with Gary Ormsby in 1989. He was part of that Funny Car team that Cruz Pedregon won the championship with in 1992, but the Ormsby championship is the one he gets the credit for. We actually got along with him earlier on, but I think when he won that championship with Ormsby in 1989, that might have led to him changing. Beard just hated my success. Hated it.

I retired from driving at the end of the 2003 season and at the NHRA Awards Ceremonies in November, I received the Don Prudhomme Award for my contributions to the sport—something I'm very proud of. After the ceremonies ended, I was walking from the Kodak Theater and saw Lee, all by himself, walking toward me. He came up to me, gave me a hug, and said how happy he was for me.

His smile was sincere, and I have to believe he meant what he said. It was a very nice gesture on his part, and I appreciated him doing it. In my mind, I felt whatever problems had existed between us were over. And I was glad. I think he was glad, too.

I'll say one more thing about him. Rahn really respects his mechanical abilities and has said more than once that Lee builds a beautiful engine and doesn't blow up his equipment. That's a strong compliment from Rahn, who doesn't pay respect to anyone who hasn't earned it.

Darrell Gwynn

On April 15, 1990, 28-year-old Darrell Gwynn, a popular star of the NHRA who had excelled in both the alcohol and Top Fuel ranks, was gravely injured at Santa Pod

Raceway in England while making an exhibition pass in his Top Fuel dragster. He miraculously survived the high-speed crash but was paralyzed from the waist down and lost part of his left arm. He remains active in the sport of NHRA drag racing as a team owner.

Shirley reflects on Darrell's accident and how she reacted to the news of it.

I had raced against Darrell many times before his accident. I liked him very much and found him to be a great sportsman. I remember we raced against him in Phoenix in 1989 and beat him. It was a great win for me, because it was the first time I had won since I had come back to the NHRA. We ran well that year with a bunch of No. 1 qualifiers and runnerups, but that was my first win of the season. Steve Gibbs later told me it was his favorite race of the entire year.

Darrell was great. After we got to the top end, he was standing up on the seat of his car, and when I looked over, he said, "Shirley, you welded me," which is drag racer slang for "You left me in the dust!"

When I found out about Darrell's accident I got on an airplane and went right over to England. I was still in some pretty nasty pain from my accident and had some bone-against-bone issues in my feet that made walking very painful. But I had to go over there. I felt I just *had* to.

When word got out I was going to England to see Darrell, a bunch of racers came to me and handed me money to give to Darrell and his family. Dickie Venables handed me five $100 bills, John Force gave me some money to take with me, Larry Morgan gave me cash, and

I can't begin to tell you how many racers came through. I had about $5,000 with me when I got on that airplane, and I put another $5,000 with it. And let me tell you, the Gwynn family was in dire straits over there.

I knew that Darrell was a big fan of orange-flavored Otter Pops. So before I left I got two boxes of them and brought them with me to the airport. I was having trouble enough walking at that time, and when we got to my departure gate, my husband, Rahn, couldn't carry the boxes down the Jetway because he wasn't a ticketed passenger. It was up to me to get them on the plane.

I don't know how I ever did it, but I carried those boxes on the plane, sat down, put them under my feet, and flew to England. And it's a good thing I had them with me because the food Darrell was getting in the hospital wasn't the greatest he had ever had. Mashed potatoes with cabbage cooked in it, things like that.

But when I first saw him, I knew he was in deep trouble. He had really been banged up, and I could tell things weren't very good.

I stayed over there for about a day and a half, and when I got back home, I got on the phone and bitched high and low to the NHRA, FIA, and said, "C'mon, you need to get some help going. There's a serious problem here."

Lynn Prudhomme took the ball and ran with it for Darrell soon thereafter. So did Sheryl Bernstein, Kenny's girlfriend at the time, who got on an airplane with Lynn and flew over there. When they did, the shit hit the fan!

Arrangements were made to fly Darrell home and even the airline came through and provided a plane that had the rear section curtained off for Darrell so he had a lot of room. Until Lynn and Sheryl got over there, nobody had gotten off their bupkis for him.

What made me feel as if I had to go over there for him? I was still in lots of pain from my accident in 1984, but someone was hurt and I felt compelled to do what I could. Why did I not trust anyone over there? Why did I think I could make a difference? I have no idea.

I just did it, and I'm glad I did. If because of what I did I saved Darrell and his family only one day of not knowing what was going on or one day of suffering over there, it was worth it.

Jimmy Nix

Jimmy Nix was another great guy. I knew him in more recent years; in fact, I match-raced against him shortly before he was killed in Dallas in 1994. Jimmy was in wonderful shape for a man his age. He looked terrific.

But I got to know his mother, Mildred, real well, even better than I knew him. We really took to each other right from the beginning, and whenever she was at the races, we'd spend time together talking. She never went to the starting line.

It was right around the time that Jimmy died that Rahn and I moved from Northridge, California, to Michigan. Our shop was right behind our house in

Northridge, but you may remember the earthquake that hit that area in 1994. That was the last straw for us. We were out of there! We sold the house and the shop and got ready to make our move to Michigan.

We loaded up a big Penske rental truck until it couldn't hold another thing. Rahn was going to drive that, and our pickup truck was going to go on a car trailer behind the rental truck. I was going to drive my Mercedes station wagon. But the pickup truck wouldn't fit on the trailer and the Mercedes would, so I drove the pickup truck and we towed the Mercedes. On the way to Michigan, the plan was to stop in Oklahoma City for Jimmy's funeral. That was going to be a heck of a trip with the amount of time we had to get there.

So there we were driving this big Penske truck that was pulling the trailer with the Mercedes while I was driving the pickup truck at speeds that we had no business driving at, trying to get to Oklahoma City in time for the service.

I don't know how we did it, but two minutes—two minutes!—before the funeral started, Rahn and I pulled up in front of the church with this whole dog-and-pony show.

We went into the church, and we gave Mildred our condolences.

It was after the funeral that we became "phone friends." She would call me before every race and after every race. If I wasn't home, she'd leave a message on my machine that said, "Don't call me back! Don't call me back! I was just thinking of you today. So don't waste your time. We'll talk later."

In 2003, during my retirement tour, I kept reminding myself to send her one of my "Last Pass" diecast dragsters. Every time I thought about doing it, for some reason I'd let it slip and I wouldn't get it done. It kept getting put off and I never took care of it.

Then, I got the phone call from Oklahoma that she had passed away.

I hated myself! I couldn't believe I hadn't sent her that gift. I was very upset that I hadn't done what I wanted to do. Mildred was a very special lady and a dear friend. I will miss her.

Graham Light

When I had the accident in Montreal in 1984, the driver's seat in my dragster was made from magnesium. I don't know if I was the only driver with a magnesium seat, but I doubt that very many other cars were fitted with one. They were very light and helped to save weight.

But when I crashed, the seat broke apart and splintered into shrapnel. We knew then and there that nobody should ever have a magnesium seat in his or her race car. We made it a point to tell the NHRA that this was something that had to be addressed.

In 1989, we were at an NHRA event when Graham Light, their vice president of racing operations came down to my pit, walked under the rope, walked into my trailer, and sat down next to me.

"I have to tell you something that will make you very happy. We've outlawed the magnesium seat," he said.

My only question was, "Why did it take five years to do it?" Maybe he expected me to give him a big hug and a big thank you, but those weren't coming. And that was what got me going on my "Graham Light Tour," if you will. I was really unimpressed with him back then. The more exalted he became the more I resented it. I just never looked at him as a knowledgeable administrator. I didn't think he had the intellectual skills necessary for that position. He just sort of fell into it because of good luck.

Carl Olson, who is a good friend and was a very capable person for the NHRA, and Steve Gibbs may not have been really big fans of Graham, but suddenly, they were taking orders from him. Again, I resented that. I felt that Carl and Steve were very well qualified, but I didn't feel the same way about Graham. There were some things that I felt Graham should have been held responsible for.

When Blaine Johnson was killed at Indy in 1996, the fact that Blaine's car hit the end of one of the openings in the guardrails when he crashed made me mad. I wrote Graham a strongly worded letter and basically read him the riot act. He never answered it.

Then when we went to the Winternationals in Pomona in 1991, we put something on the car that wasn't my idea personally, but I condoned it. A caricature was painted on the wing of the car that showed Graham's head that looked just like him with his white hair and all, and there was a red circle around it with a red slash going through it.

The NHRA tech inspectors in the tech line saw this, and they were going bananas. When they took a look at this little cartoon, they went crazy. After one of our qualifying passes on Friday, we were back in the pits. Steve Gibbs was walking toward us. He looked at my husband, Rahn, and pointed to the cartoon on the wing.

"Rahn, you have to take that off the wing," he said.

"I'm not taking that off the wing. It's a free country, and that's just free speech," Rahn said.

"You have to take that off the car," Gibbs repeated.

"Well, maybe I'll just take the car over to the Media Center and tell them I've been told I have to take it off the car," Rahn retorted.

"The Media Center is right over there, but if you don't take that off the car, you're not going to the line," Gibbs stated.

Well, we didn't take it off the car, went up, and made our next run. When we got back to the pits, low and behold, here comes the president of the NHRA, Dallas Gardner!

Dallas walked over to me and said, "Shirley, you have to do me a favor."

Just about anytime he came to me to talk to me, that was how he opened the conversation: "Shirley, you have to do me a favor."

I said, "What's that?"

He said, "You have to take that off the car."

"No," I said, "I'm not taking it off the car. Dallas, here it is the first race of the season, the Winternationals, and this is the most important thing the president of the

Rahn has always watched every move around the dragster, and he is not afraid to call out other drivers or the NHRA on things he believes are handled incorrectly. *Shirley Muldowney Collection*

NHRA has to do?! You have to come over and talk to us about that dumb cartoon on our car?"

He said, "It's a personal attack."

Well, finally we did take it off the car and replaced it with a picture of Bozo the Clown, with the big nose and the orange hair with the red slash going through him.

A little later, Gibbs came back over.

"Don't you have any respect for the president of the NHRA?" he asked Rahn.

"I have as much respect for the president of the NHRA as I have for Bozo the Clown," Rahn said as he pointed to the wing.

An Apology

Ever since I came back at the U.S. Nationals at Indy in 2001, Graham Light has been great to me. Just great.

I've sat in on drivers' meetings, which they didn't have a while back like they have today, and I've found him to be a very capable administrator. I'm not going to say he's up to speed on the best way to build a chassis or what's the best combination to run in those cars—maybe he is—but when it comes to moving us in the right direction and organizing what needs to be done in the areas of safety, he's got my vote.

At the 2003 NHRA Awards Ceremonies I gave my speech after receiving the Don Prudhomme Award and I did apologize to Graham Light. In fact, I said that I owed him the biggest apology of all.

Bill Bader

Bill Bader runs the IHRA, and I don't like him. For my money, he's a son of a bitch. Rahn asked me why I would even mention him in the book, but I'm going to.

In 2001 when we didn't have a big enough budget to run more than a couple of races with the NHRA, Fred Wagenhals of Action Collectibles came to us with just under a half million dollars to run eight events with Bader. We figured if nothing else it would be a pretty good test and tune.

So we headed up to Grand Bend, Ontario, and there were nine cars total there. Bobby Rex's car crashed really badly, and the driver, Doug Foxworth, was hurt. With back injuries and head injuries, he was in the hospital for a while. At that point, he had been the No. 2 qualifier.

Bader decided to take a car from a team we had never even heard of—a car that couldn't even make it down the racetrack—and inserted it into the No. 2 spot! And Rahn went ballistic!

Rahn gathered all the drivers together in the staging lanes on Sunday morning prior to the race and said, "Now's the time that we bring this man to his knees."

What Bader was trying to do was terrible, and Rahn didn't want anything to do with it.

Doug Herbert said he was with us, but Jim Bailey, Bruce Litton, and the other drivers, who when it came to Bader had no backbone, began to put their helmets on. It was terrible that Bader was going to get away with what he did. We said then and we said until we retired that we would never race with Bader again for any event, any amount of money, or any reason.

Because of what he did then and so many other things he's done over the years, I think he was the worst thing to happen to drag racing.

Ron Colson and the Phone Booth

Ron Colson was a Funny Car driver back in the 1980s, who thought he was a much better driver than he really was. He drove the *Chi-Town Hustler*, the famous

Funny Car based out of Chicago, and that was his biggest claim to fame. He went on to do some public relations work in the sport for a while, just sort of bouncing from place to place.

He wound up doing PR for Bill Bader at the IHRA, and I can remember I was booked into a couple of IHRA races and was asked to go on some promotional junkets to help sell the event. One of those races was out at Cordova, Illinois, and the other was in Stanton, Michigan, a track where I had run very, very well in the past.

I would meet Ron in the morning, and he would pick me up in his car—if you could call it a car—because it was pitiful. He picked me up for the junket in Cordova in a car that had a crack in the windshield that stretched from my side of the car all the way over to his!

When I saw this, I didn't even want to get in! Later I told the management at Cordova that they needed to buy him a car, loan him a car, rent a car for him—whatever they needed to do—because this was embarrassing! Especially when TV reporters would come out to the car to interview me after I had arrived in a real crapcan!

When I was out at Stanton, I decided that we'd head over from Stanton to Kalamazoo and Grand Rapids and anywhere else we needed to go.

I said, "Ron, today I think I'm going to take my own car."

He said, "Really? How come?"

I said, "Well, because I want to visit some people when we're done, and you can go you're way and I can go mine."

He said, "That's fine."

It was one of those days when you could tell he really didn't have his act together—not that he ever did, but that day it was very obvious he was playing it by ear. I noticed that every so often, he'd have to stop at a phone booth and make a call. Some of the calls would take several minutes, and he kept pulling over to make them. It was becoming a real nuisance.

Finally, he stopped to make another call. It occurred to me that he might be setting up interviews as we were going from place to place because he hadn't made any of the arrangements ahead of time. That's when I had to put on a little hissy fit and say, "C'mon, what are you doing? We're going to be late!" By now, I was really fed up.

He was still in this phone booth with his back to me and didn't turn around. He just kept talking to whomever he was talking to. And as he did, he took his hand and sort of waved me off as if to say, "Cool it!"

That was enough for me. I turned around, got into my car, and drove off with him still on the phone. That was the end of that promotional junket! In fact, when he got to the next interview, they interviewed him! Whenever PR people hear that story, they get a big laugh out of it. And I'll tell you this: It's a good thing he had a car because if he hadn't had one, I still would have driven away and left him on the phone.

He should have known that I wasn't the type of person he could just wave off like that. Maybe he was trying to flex his muscles a little, I don't know. But I don't have a lot of patience with people like that.

"Goodbye!... See ya!"

Scott Geoffrion

I never knew Scott Geoffrion very well. Before last year, I might have said hello to him once or twice, but nothing more than that. I had heard from more than one person that Scott was quite the partygoer, especially when he drove for Mopar. Scott's a good-looking guy, certainly not a bad driver in the overall picture, but probably never was in complete control when he drank.

At Indy in 2003, qualifying had ended for the day on Saturday, and there was a bunch of people hanging around in my pits. The guys were working on the race car, we were getting some food ready, and we were all enjoying ourselves. At one point, I looked over and saw Scott sitting on the other side of the race car in one of our chairs—not very far from where I was—with his girlfriend on his lap. But she wasn't just sitting there; she was practically giving him a lap dance!

There were kids in the area while these two were kissing each other, groping, and carrying on like you wouldn't believe. I watched what they were doing and put up with it for about 10 minutes.

Then, I walked over to him, tapped him on the shoulder, and said, "Scott, you're gonna have to take this someplace else."

Scott looked up at me and said, "What?!"

I said, "You're gonna have to take this someplace else. Now!"

He said, "Shir… Shirley, hey, uh, you know me! You know who I am, right?"

I said, "I know exactly who you are, but you're gonna have to leave, OK?"

Then he got a little defiant. He was trying to make a fool out of me. Then he started laughing.

At that point, I said, "Out!"

Nothing.

"Look, you've got 30 seconds to get out of here, or I'll call the cops and they can get you to leave."

Scott stood up and looked at me, and my husband, and all these people standing around and started dropping the f-bomb on all of them. "F--- you!" "F--- you, too!" and so forth. I had some people there from my sponsor, Action Collectibles, and some other guests so this wasn't a very good thing that was happening.

I walked back over to him and said, "What did you say?"

And when I said that, he took both hands and he pushed me. Of course, with me being very topsy-turvy with one leg shorter than the other, I went down. And I had this black handprint from where he pushed me right there on my neck.

It wasn't more than a split-second later when Donnie Couch had Scott down on the ground. Donnie was just about to pop him, but he held back, and I give him credit for showing some restraint and not doing it. He controlled himself, and then let Scott get back up.

When Scott tried to get up, he was so tipsy that he fell over facefirst, right there in front of everybody, and cut his forehead. That's what really happened.

Rahn would always tell me, "Honey, don't ever go back like that and get in someone's face. Don't go walking over to someone whose giving you trouble."

Donnie had his own opinion: "Shirley, if you were a man you'd be dead by now."

Later on, the police came by and asked us what happened.

When we told them what Donnie had done because of the trouble Scott had started, they said, "That's all we wanted to know and that's OK with us."

I think the police had already seen the condition Scott was in because he had told them he had been accosted and wanted them to do something about it. He was clearly over the top. The police didn't arrest Scott, but I think they did gather him up and take him away so that the situation could be put to rest.

The next day, Graham Light of the NHRA came down to my pit and said, "Shirley, I heard you had a problem down here."

And I said, "Yes, but it's OK, and it's all taken care of."

He said, "It's not OK, and I want to know what happened."

So I told him what had happened, and he said something needed to be done. Scott was out of control.

In fact, at the next race in Reading, Pennsylvania, we heard Scott got into trouble again when he got into a barroom brawl and threw a full glass of beer in someone's face. In my opinion, Scott needs to get some professional help or he'll always be in trouble.

Revenge?

I never expected Scott Geoffrion to apologize to me, but I told him he should apologize to Donnie because Scott's version of the story claimed Donnie Couch had hit him. That's not true. There were plenty of witnesses standing there when it happened.

But I heard that Scott went to Roy Hill, who has been know to help people settle a score or two, and set it up to have Donnie taken out. As you can imagine, Donnie was very upset when he heard that. Donnie had gotten a call from some people who had worked for Roy and told him what was going down.

They said, "You need to know that Scott is setting it up with Roy to have you taken out."

So I called Roy Hill on the phone myself. Roy denied that he had planned anything with Scott. I was sitting in my car with Donnie and Pat Galvin, and I told him that we had gotten it from "almost the horse's mouth."

Roy hemmed and hawed and said, "C'mon Shirley, we've been friends for a long time. ..."

I said, "This isn't about who's friends with who. This is about what we heard. Don't you ever think about coming after Donnie Couch."

I told him that if it ever happened, we would take legal measures. And that's when it ended, right there.

Some Neighborly Advice

It was a couple of races later that I was going into my hotel room after I had been out on a shopping run. I stepped into my room and as I looked down the hall, I saw someone coming. My door was open, and I looked out as he walked by. It was Scott Geoffrion.

I saw him slip the key into the lock of the room right next to mine.

I said, "Scott?"

He turned and saw me and said, "Shirley!"

He walked over to me and gave me a big hug.

I said, "Scott, I want to talk to you. Take it from someone who knows. You're headed for trouble."

We had a nice talk in the hallway, and I tried to get him to see he was jeopardizing his career. He thanked me and gave me another hug and I could tell he had had just a drink that night—maybe one drink.

He couldn't have been any nicer to me that night, but I can't help but feel my advice went in one ear and out the other.

More Television

I have been where no drag racer has ever been. I was the first drag racer to be invited for a ride in a navy F-18. I was the first and only drag racer to be invited on *The Tonight Show*. I know, I'm bragging now. But it goes on and on. I was the only drag racer to appear on ABC's *World News Sunday Night*. I was on *Good Morning,*

America. I was interviewed by Maria Shriver, David Hartman—to name a few.

In 2003 I was on Jim Rome's sports talk show, and afterward, he said mine was one of the best interviews he had ever had. He had Evil Knievel on his show around that time and someone e-mailed Jim and said, "You should have Shirley and Evil on the show together. The two of them could probably end the war!"

I have all the tapes of all those shows.

I've done everything I've ever wanted to do—and much more. A whole lot more.

Not only have I been on television and met many famous people, I have met a lot of celebrities who are drag racing fans. In 2001, Richard Dreyfuss visited us at U.S. Nationals. *Shirley Muldowney Collection*

Ron Lewis Photography

Chapter 4

The Leading Lady

With 18 national event wins and three championship trophies in 1984, I kept my motivation to win. *Taro Yamasaki/People*

Motivation

In the early days, I had no doubts that a lot of people I raced against hated me. You had a bunch of spoiled brats, a bunch of farmers, good old boys—you had all kinds of types out there. And not only then, but throughout my entire career, that's what motivated me to win.

Sorry, Not Today

One time in 1965 Jack and I towed all the way to Bristol, Tennessee, because we heard there would be some openings on the entry list at the event down there. I remember being one of several people standing on the sidelines waiting for the NHRA to decide if they would open the field and allow more entries into the race. I had been sitting on the ground for a while outside that little wooden booth that passed for a tower just waiting to find out if they were going to allow me to race.

Inside, Eileen Daniels, the wife of longtime NHRA Division 3 director Bob Daniels, was going to make the decision, and the NHRA wives could be as conniving as the husbands. So she took great pride in informing me I wasn't going to be part of that race.

I decided I'd try to find Wally Parks, who I knew was staying somewhere in town and plead my case to him. I finally found the Holiday Inn where he was staying. When I saw him, he looked tall and perfectly tanned and was wearing a light-colored suit, which made him look like a brand new penny. I told him what had happened at the track.

All he would say to me was, "I'm sorry, but whether you're allowed to race or not isn't my decision to make."

That was that.

I remember those days. Over the years, Eileen and I have been friendly. In fact, after I was badly hurt in Montreal in 1984, I was undergoing surgery in Indianapolis and Eileen made it possible for Rahn Tobler, my crew chief, to stay in a hotel nearby for only $20 a day.

But sometimes I wonder if some of the people who made it so difficult for me have any memory at all. I'll never forget that day in Bristol, but as time passed, they treated me with more respect. Once they learned I wasn't just a flash in the pan and that I could drive—and put fans in the grandstands—they came around. But they did everything they could in the beginning to keep me from doing what I wanted to do, and that was race.

Shirley's Top Fuel License

Here's the truth on how I got my first Top Fuel license. I had had a number of fires in the Funny Cars I raced, and Poncho Rendon was a well-known racer in the Detroit area and a friend of Connie Kalitta's. Rendon was a consummate gentleman and a real good guy. In fact in 1973, when I was having trouble finding someone to work on my Funny Car for me, he did.

Poncho owned a Top Fuel car back then, and he had a driver named Gene Domagowski, whose dad had a tool and die shop up around Frazier, Michigan, and so Gene was one of the gang. He was kind of a quiet guy but everyone liked him.

Unfortunately, Gene drove all night going to an AHRA race in St. Louis one weekend and was really tired. Poncho noticed Gene was closing his eyes and resting as he sat in the race car before he fired up the engine.

Poncho saw him and said to him, "Don't do that on the starting line!"

Gene pulled to the line and was very late at the starting line. He must have been looking over at the other lane when the front wheel of his dragster went off the track, which had a dangerous dropoff into some dirt. It dug in at about 200 miles an hour and snapped the car over, which meant it may have been flipping over at 300 miles an hour. Gene was killed.

But before that accident Poncho asked me if I'd match race his car, and to do that, I'd have to upgrade my license. So Poncho, Connie, his son Scott, my son John, and I went to Cayuga with Poncho's dragster. "Big Daddy" Don Garlits was there that day, and he was match-racing Tommy Ivo.

That was the day I realized I had to sit up straight in the car to gather myself. After Connie helped me brace my helmet inside the rollcage, I went out and made two passes at around 219 miles an hour. That was fast enough to get my license upgraded, because I had already run 220 miles an hour in the Funny Car. Garlits signed my upgrade as did Connie and Ivo, and it was issued. It was as simple as that.

Here I stand after I received my Top Fuel license in 1973.
Shirley Muldowney Collection

There really wasn't any fuss to get them to sign my Top Fuel license, unlike the way it was portrayed in the movie *Heart Like a Wheel*. Soon thereafter, I went out and was match-racing Poncho's Top Fuel dragster. He lined up several dates, and we made some pretty good money. The asking price for us to come and run was $750, which wasn't bad money at the time.

But when I went for my first NHRA license in the B/Gas dragster at Connecticut Dragway in 1965, I did have to use some persuasion to get that license signed. But the movie couldn't get into the whole business of two licensing procedures, so that's why it was written the way it was. Something else that was in the movie that didn't really happen was that Wally Parks wasn't in attendance when I received that first license.

After I got my license at Connecticut Dragway, it was submitted to Darwin Doll, the NHRA Division 1 director, for final approval. Now those division directors, I tell you, were like the mafia. They were quite a bunch. The old regime. Pains in the asses—all of them.

Well, Darwin called the owner of Connecticut Dragway, Frank Marotta, and asked what he thought about me getting the license.

Frank said, "She can drive!"

That should have been the end of the story, but it wasn't.

A week later, I was sitting at my desk at the job I had at the time—I was a receptionist in the display advertising department at the *Albany Times-Union* newspaper. I received a letter from the NHRA, which said there was no way they were going to honor the license that was issued to me at Connecticut Dragway. It said Darwin Doll had made a mistake. It went on to say that women don't drive dragsters; women only drive stock cars.

And I said to myself, "Oh yeah?"

The *Albany Times-Union* had just run a feature in the Sunday edition's pull-out supplement about me with my picture on the cover, and the newspaper jumped right on that letter from the NHRA. They were all over it.

They ran a story that said the NHRA was discriminating against women and it was a disgrace. Well, Wally Parks must have gotten together with his lawyers because a week later he called me at the newspaper and said, "We'll honor your license ... locally."

More Hurdles

I had an NHRA license, but whenever I wanted to enter a national event, I'd get a letter from Jack Hart, the NHRA's technical director that said, "You've never done anything that would entitle you to compete at a race as prestigious as an NHRA national event. Your application to race is denied."

They just shined me off as far as national events were concerned. I could run at the local races, but that was it.

It went on like that for two years. During that time I'd pick up the phone and call the NHRA and rant and rave at all different hours. One night I called and there was a guy working there late, and he just happened to pick up the phone. It was Steve Evans; I forget what his capacity was at the NHRA then.

But he said, "We'll accept your application to race at Englishtown. Just show up."

And I did. And I just barely missed qualifying for the field.

Another Attempt to Break the Gender Barrier

After I missed making the field in Englishtown when I finally got to a national event, it was a while before I entered another one. It wasn't until the U.S. Nationals in Indianapolis in 1970 that I gave it another shot with our twin-engined dragster. I missed the field that year by a couple of 1/100s. That was the first time that Wally Parks ever saw me drive.

But I thought the twin-engined dragster was cool. I liked it. It had two blown and injected Chevy engines. In 1968, my son, John, was 10 years old, and he and I drove by ourselves to Illinois to pick up a trailer and then to Gardena, California, to pick the dragster up at Don Long's chassis shop. In fact, at that time we had an AA/Gas dragster and a BB/Gas dragster before I ever met Connie Kalitta. So anyone who says I hadn't done anything before I was with Connie is just not correct. Jack Muldowney did a great job.

We had run a lot of races, and I wouldn't say we didn't get noticed, but in the eyes of the NHRA back then, if you weren't from California, you didn't count.

My dual-engine dragster was able to clock speeds of 190-plus miles an hour. *Shirley Muldowney Collection*

I Am Woman

Other women had been drag racing before I finally went into Top Fuel. You can't overlook the fact that other women tried to do what I did. You can't take that away from them. But I didn't want to race just to be able to say that I could do it. I wanted to win, just as badly as the men did. Sometimes more.

The one thing that was always important to me was not to be given any special treatment because I was a woman. Subconsciously, I always felt that if I were to be given any kind of favoritism that that would be a disadvantage. I had to overcome the possibility that I could be given an easier ride than someone else, which to me would have been just as bad as being discriminated against.

The things I take a lot of pride in are the things we did, not so much at the national event level at the good tracks and the safe tracks, but the races we ran at some of the smaller, lesser known tracks. The match races, sometimes at shaky facilities and sometimes at night, were the times when it was "do or die" whenever you came to the starting line. Those are the things I wish a lot of people could have seen. It was a lot different from what racers have to deal with today.

Other Female Racers

A lot of people have said I never liked other female race drivers, but that was never true. I liked all the other women drivers, even though I was accused of hating them.

There's not another woman driver I ever hated. I didn't hate Lori Johns. She was a product of the times, and I think there were more people responsible for her being disliked than anything she was responsible for that made her disliked. She had people around her who hated me and just fanned the flames that erupted between us. It was perfect for them.

What was really a shame was that when all that bitterness was going on between us, we were struggling. That allowed those who didn't like us to get even. But in the end, we came out on top.

Aggie Hendricks was a jet car driver, and I thought she had the right stuff. I really liked her.

I remember I match-raced her in Port Charles, Louisiana, one year, and that track absolutely packed the grandstands for that event. I don't remember who won or who lost, but I do recall that her jet car ran some very big speeds—but it took forever for it to get there. We had dinner with her and her team, and we enjoyed each other's company.

Sheila Tafur-Kopchick is a lady who could do very well in the right Top Fuel car. For a while she handled all my souvenirs when I was out doing match races.

Her husband, Mike Kopchick, who owns Rage Racing, was on our crew, and he met her at Lebanon Valley, New York, several years ago when she came out to see me. They are the perfect couple.

Sheila drives an A/Fuel dragster, and she learned a lot of what she knows about driving a race car from the time she spent with me. I let her warmup my dragster on occasion, and we did an awful lot of talking about racing.

Sheila got as close to the way I go about driving a race car as anyone out there. I always wanted to see what she could have done in a Top Fuel car that was set up right.

She still races her A/Fuel car, mostly in the Northeast, but I really believe that if she were to get the chance to drive a Top Fuel car that was set up properly, she could get the job done. She's got that fire. Absolutely.

Lori Johns

The problem I've always had with many of the women who have come into Top Fuel is that they make a few runs and suddenly think, "That's all there is to it." There's a lot to know about racing in a Top Fuel dragster, and most everything you have to learn comes from experience.

Lori Johns came into Top Fuel because her father was able to buy her whatever she needed to go racing. Don't get me wrong, it wasn't like he was the CEO of Microsoft. He sold used cars. That's where he made his money.

In the beginning, she was able to get down the racetrack, and she had Larry Meyer working on her car, which produced some good numbers. So she seemed to think she knew it all.

At the first race where we were both running, she just popped right into my trailer and acted like it was Old Home Week, and I tried to tell her it wasn't all as easy as it looked out there. Of course, her daddy was paying for everything, which isn't the way that I did it when I started out. But I understood that every racer who came on the scene wasn't going to do it the way I did it.

Before she moved into Top Fuel, she had the accident in her alcohol car that led to the lawsuit she filed against the racer who collided with her. At that time, I was recovering from my accident in Montreal, and I was still having a problem getting around. I could barely walk.

Scott Kalitta and a few of the other kids in that age group came to me while we were racing in Baton Rouge and wanted me to go visit Lori, who was in the hospital. My son, John, came with me, and we found the hospital down in the center of Baton Rouge. Lori's father was very happy to see me when I got there, and Lori didn't appear to be hurt that badly. As it turned out, I was told she had sustained some neck injuries. She was less than happy to see me.

I didn't spend much time there, but I did offer her my best wishes. That was it.

She moved on after she recovered and went to the Frank Hawley Drag Racing School to prepare to come onto the Top Fuel scene. But I got the feeling that when she moved onto the scene that she'd do whatever she could to use my resources. By that I mean, she had this lawsuit baggage, and it would have been really good for her to get me over on her side. I wasn't going for that.

I always got the feeling she had an air about her and pictured herself as high society. That just didn't play well with me. It was almost laughable.

But her lawsuit is really what upset me. And a lot of my displeasure was with the NHRA and Dallas Gardner, who was the president of the NHRA then. (I was never a big fan of his, and I'm still not. I always felt he's taken a considerable amount from the sport, and I'm not sure whether he's ever given anything back. I really can't put my finger on anything he's done to help the sport, but I'd be very open to hearing something I might have missed.) But it's my feeling that it was Dallas Gardner's decision to allow her to race after filing that lawsuit.

I didn't race Lori a lot, but it seemed that every time we weren't running well, there she was in the other lane. And it was usually at a time when she was running well. For those who didn't like me, she was a dream come true. And of course, she played right into it.

It was like throwing salt into the wound, and of course, I resented it.

I'll never forget the two years in a row that I didn't qualify at Indy. That second year, 1989, my husband, Rahn, had sent me to get some food for the guys during a rainstorm that came in. We were in the field at that point, but while I was gone, the rain stopped, the sun came out, and they dried the track.

The track was prepared and not only had the racing surface gotten better, but the air was drier and the conditions had improved from the earlier sessions. The result: We kept getting bumped down, and then bumped down again, and finally we didn't qualify. Lori took great pleasure in that. You could tell by her emotions and by how she carried herself. Believe me, she got her licks in too.

We left Indy and went home to get ready for Maple Grove. We heard a knock on the door, and it was a deliveryman with a box of flowers. We opened it, and there was a bunch of roses—not the expensive long-stemmed roses but cheaper ones—that were painted black. There was a card with the roses that said, "You have finally faced reality."

I went out into the garage and blew my stack! I threw things all over the garage, and all the guys got as angry as I did. Of course, I knew who sent those roses.

I had done business with the flower shop that delivered the roses for 25 years, so I went over there and begged them to tell me where the flowers had come from.

They said, "You know, we didn't like taking this order when it came in. We didn't like taking those roses and painting them black."

It took a while for them to tell me but they finally did. The flowers had been sent from Corpus Christi, Texas. That's where Lori lived.

Some years after that, we were told that Lee Beard had something to do with it. Everyone knew he and Lori were friendly, so it was obvious to us that he had a part in it.

When we got to Maple Grove, we were ready to get down to business. On our first pass off the trailer, we ran the quickest run in NHRA history! I was the first woman in the fours, and Lori went nuts in the pits, crying and carrying on like a baby. Meanwhile, out at the starting line, Darren Capps, who was on my crew, looked at the scoreboard as I crossed the finish line and with both hands flipped the bird toward the tower! He was sorry later because the tower was filled with people visiting from Japan, but at that moment, he couldn't help himself.

The next day, on Saturday, when we got to the pits, there were roses from fans, racers, racers' wives—all over our trailer. They were all thrilled. I wasn't the only one who was sick of her.

On the next run, I raced Joe Amato and he ran a 4.96 while I ran a 4.97—the first side-by-side four-second runs in history. It was a big day for us and a day of reckoning for her.

But as people get older, they usually move on and change for the better. I believe that and I believe that's what Lori did. But when she came into the sport, she got a little glory and thought she was the whole story.

One final story. In 1990, Lori's father brought my dog at the time, Skippy, back to my pit area in the dead of night after we finished qualifying. Skippy was 17 years old then and just about blind, and Lori's dad knew she was my dog.

When he brought her back, he said, "I like dogs."

I'll never forget that. Whatever issues I may have had with Lori didn't extend to her father.

Lucille Lee

Lucille Lee was a Top Fuel driver I really liked. She was always sincere. She was real. That's what I picked up from her right from the start. But she drove for a guy who I really didn't care for at all. A lot of people agreed with me. He was considered to be a very bad character. His name was Mark Danekas. He had done time in prison and was a bad apple. His business associate was Ronnie Capps, the cousin of Darren Capps who worked for us, and Darren had no use for him. Ronnie was another loser. He was a bosom buddy of Lee Beard over the years and worked with Beard on Kenny Bernstein's team back in the late 1990s.

Danekas owned the car Lucille drove—at least I *think* he owned it, and they had kind of a shaky sponsorship deal with TR3 Resin Glaze. I remember in 1982, Lucille beat us in the final in Bakersfield, and we had a serious problem with the clutch. The clutch hadn't blown up, but it was really in a bad way. It was obvious we had some difficulties with it.

When Lucille got down to the other end, she was asked what it felt like to beat Shirley Muldowney.

She said, "I didn't beat Shirley Muldowney. I won because my car ran better than hers. She had a problem, and I got there first."

I remember one other thing about that race. As I was rolling down the track with the bad clutch, I was taking the car to the far turnoff. I didn't take the short turnoff. I could face the music when we lost, and I was going to the far end.

As the car was making its way with all of this grinding and crunching coming from the bellhousing, I looked over and saw this guy jumping up and celebrating—and I mean really jumping up and down! His feet were four feet off the ground! And I'm watching this guy jumping way up off the ground and thinking, "Who *is* this asshole?"

It was Dale Armstrong.

I had never had any issues with Dale, but for some reason, he was not a big supporter of me. He was tuning for Kenny Bernstein that year, and they didn't qualify for that race. Maybe he was getting some satisfaction out of seeing me lose. I don't know, because I really didn't know him at the time, although I knew who he was.

A few days later, Lucille told me that after the final round, she wanted to come over to me and shake my hand, but Dale had said to her, "Don't go over there. You don't want to go over to her."

And so she didn't.

I never liked him after that. In 1986, I was interviewed for a newspaper article while we were in Brainerd, Minnesota, that was being written by a guy who was always looking for a controversial angle in his stories. Somehow, Dale Armstrong's name came up.

I looked this writer in the eye and said, "Dale who?"

Dale was part of that old "clan"—the guys like Dave Settles who became the enemy. I knew who the enemy was, believe me. I knew exactly who they were because I always wanted to know who was who and where I stood with everyone out there.

Janet Guthrie

I always thought Janet Guthrie was a goof. The bad feelings between us started when I was at a press conference the same year she first came to the Indianapolis 500.

Bob Daniels stood up and asked me, "What do you think of the fact that Janet Guthrie thinks you're nothing more than a six-second driver?"

I said, "Really? She should come out to one of our races and see a real race car."

My comeback could have probably been better, but I was caught off guard. I'll tell you this: I was chapped for the longest time after I heard that. Then later on I began to wonder if maybe I hadn't been set up and reeled in.

Regardless, after that happened I did make some comments about her in a couple of newspapers such as, "She's the only driver in the Indy 500 who doesn't need a rearview mirror," or "I'll give her some credit when I see her tap on A.J. Foyt's bumper to get him to move over." I never saw her fight for position. Ever. If she wanted my respect, I would have needed to see that. I never did.

Lynn St. James

Lynn St. James was another female race driver who never seemed to be able to fight for the position ahead of her. I thought she was in the way. I don't hold anything against her personally, other than I think she's a much better talker than a driver.

She used to talk all the time about performance and how important it was. Over and over she would talk about her performance and so on and so forth. Her performances never won races. Time and time again she would ask me to endorse her programs off the track, and I would say, "No, I can't let you use my name."

I listened to an interview that she gave on the radio in Detroit a month or so ago, and I was very impressed by how great a talker she is. But as a racer, I was never impressed.

Shelly Anderson

Shelly was good. I think she did OK in her car, but I never really talked to her much after she started driving in Top Fuel. She was like other female drivers who kind of get into their own little world after they've made a few runs and say, "This is easy!"

The mistake she made was when we made our comeback to the NHRA in Dallas in 1997, and she was in her full bloom. There was an article in *USA Today* that had quoted her saying something like, "It's nice that Shirley is finally coming back to run with the NHRA because the last time she did, she didn't qualify, and who would want to end their career like that?"

After I saw that, I sent her a postcard and told her, "I never said I was ending my career, Shelly. Sorry. That must be wishful thinking."

That's all it said, and I never heard anything back.

Another thing that involved Shelly really torqued me off. When she was working on the ESPN telecasts several years ago, Dave McClelland would introduce her as a "five-time NHRA champion." Shelly won five national events, not five championships. I wrote Dave a note and pointed that out to him.

But Shelly could have spoken up before I did and told him, "I'm not a five-time champion." She was guilty of that, but she probably had a lot going on at the time, and how she was introduced was the least of her concerns.

I just felt it was disrespectful.

Angelle Sampey

God knows I wouldn't get on a motorcycle. I'm much smarter than that. I couldn't even ride a two-wheeler! So to say that I don't give her the utmost respect and admiration simply isn't true. I think what she does is a wonderful accomplishment. To race that way and look the way she does in leather with that petite figure is really impressive. Angelle is one of the prettiest girl I've ever seen out there. She's a doll.

I thought Angelle was disrespectful. When she was zeroing in on my record for most career wins by a woman, it bothered me, and a number of the wives of Top Fuel drivers such as Ally Dixon, said that Angelle didn't make a more obvious distinction between racing a motorcycle and racing a Top Fuel dragster.

But she should have made mention for the masses that she rides a four-cylinder gasoline-burning motorcycle, and it's nothing like a nitromethane-burning Top Fuel dragster, which is what our sport is based on. Motorcycles didn't even have a professional class until Winston laid down the law to the NHRA and told them motorcycles were going to be a professional class. That opened the door for another woman to come in and win a championship, and Angelle was who they wanted to do it.

One more thing. I was at Indy in 2000 and heard Dave Rieff of ESPN, doing his usual job, which is very professional and as good as anyone. He started in with the Angelle accolades on the TV coverage, and later came over to my pit. As he was walking in, I stopped him in his tracks. I took him by the shirt and brought him over to my dragster.

Pointing to my car, I said, "David, does *that* look like a motorcycle?!"

That's all I said.

He replied, "No, Shirley, that does not."

End of conversation.

But here's my only real complaint about Angelle. I just don't want to go to church every time she's interviewed. She can go to church any time she wants, and I can go whenever I want, but I don't need to hear a sermon every time she crosses the finish line.

Louise Smith

Louise Smith is a sweetheart. A very special lady. I had always heard of her but never got to meet her until I did an auto show back in the mid-1990s. She was there, and we talked quite a bit. If you're not familiar with her, she was a NASCAR driver back in the early days of stock car racing, and it was very flattering to hear her tell me what a big fan of mine she was.

She was at the Hall of Fame dinner last year, and my husband, Rahn and I just loved hearing her stories. I'm told that "Big Bill" France thought the world of her, and I can see why. In some ways, she did in NASCAR what I was able to do in drag racing, but she didn't have the same kind of exposure that I enjoyed. Knowing her today at age 93, I can tell she probably was much more ladylike during her career than I was during mine.

She has every bit of my respect.

Louise Smith (right) was the first female inductee into the Motorsports Hall of Fame. I have always admired her and I love hearing her stories. *Shirley Muldowney Collection*

DRAW

Don Prudhomme's wife, Lynn, is a very, very smart woman. She was responsible for starting DRAW, the Drag Racing Association of Women. There were a lot of other women who were there in the beginning, too, and I could go on and on, but Lynn was really the one who should get most of the credit for being the foundation of the whole thing. DRAW has come through with help for a lot of racers and their families who have had accidents while competing.

Thanks to DRAW, I got some very much needed help when I had my accident in Montreal. They do a fantastic job.

DRAW doesn't get as much support from some of the professional racers as it should because of some displeasure with what DRAW has done outside of the pro categories. I became a little disenchanted with them when they offered some help to people who were doing some stupid things—grudge racers, mini-bikers who were out at the starting line running into each other, and things like that. They were doing things that they had no business doing, and DRAW was shelling out thousands of dollars to them.

These individuals were getting hurt and I understand that, but I think that using the resources of DRAW in those instances had something to do with some pro racers not getting behind DRAW as much as they did in the beginning. DRAW still does tremendous work for people who need help, but the concept of the organization, to my understanding, was for professional drivers.

Regrets

Sure, if I could go back, there might be a person or two whom I owe an apology. I talked about Graham Light, and we've worked out our differences. There are many people who I see today and have accepted a "Hi, how are you?" from them, and that's good enough for me.

I can tell you something I am truly grateful for.

The things I *don't* know.

There are people who I know have run me down in the past, but if I talk to them today, I just let it go. I don't want to hold a grudge or live in the past, so it's better to get over it and move on. That's really how I feel.

CB Chatter

When we were on the highway driving from place to place for so many years, we'd sometimes listen to the CB radio in the truck. A few times I wanted to get on there and get into it with some of the truckers we listened to, but my husband, Rahn, always told me he didn't think it was a good idea, so I didn't.

But the truckers would see our tractor-trailer go by with my name on it, and we'd hear things like, "I remember when I saw her race in 1972, and that was the coolest thing I'd ever seen! She changed my life!" Or, "Before my wife died she was Shirley Muldowney's biggest fan!" and things like that. You have no idea what wonderful things those truckers would say.

Sometimes they'd ask on the CB, "Where's Shirley at? Is she flying in her jet?" And Rahn would answer, "No, she's sitting right here over in the right seat." And as we passed some of them, I'd wave to them, and they got such a kick out of that.

But the most memorable thing I ever heard on the CB happened not too long ago. We were driving along on the turnpike in a long line of trucks, and we pulled out to pass them. We had our CB on, and as we started going by, I heard a truck driver come on and he said, "Hey, y'all… look up. There's history going by."

That really touched me. It made me feel special.

Doing It My Way

When we were winning, we were dominant. I was strong. I was a fighter. I'd go after anyone and then go back at them if I had to. That was my makeup. When I was in the heat of battle, I fought with every ounce of strength I had. Not every driver was like that, and not many women—even today—take that approach.

I've always told it like it is. I don't know any other way, and there were times when all I could think of was, "Let me at 'em!" I probably would have gone much further in my career if I hadn't been that way. If you look at what I did in terms of today's day and age, it wouldn't work now.

But I am what I am.

Then Versus Now

I have often said that I got out of the cockpit at the right time. The sport has really changed, especially over the last few years, and I felt I timed my retirement perfectly. There are a lot of new rules, new people, and new business interests that have changed the way things are done out there, and I'm not sure that they have all been changes for the better.

Something that I take a lot of satisfaction in is that when I quit, we were running quicker and faster than we ever had before. I think some people stayed out there a bit too long, and they got to the point of undoing all of the great performances they had accumulated in their prime.

But I never grew tired of the driving part. I still love the way a nitro engine sounds when it's idling. And I always loved the way the car pushed you back into the seat when it launched. But the business part of it, the politics, the turf battles with the sanctioning bodies, and some of the attitudes of the new generation of drivers who haven't had to fight a little to get an opportunity to race—those have been the things that I found tiresome. I was ready to move on.

A Day in the Life

Today, I'm really enjoying life. I'm sitting here in my house in Willis, Michigan, on my five-acre lot, looking out at the forest, which looks beautiful as the foliage turns into all those colors. I can sit back and enjoy it.

I've got my dogs, Peanut and Amy, and my birds and wild kitty cats with me. When I have some time between my appointments, I really enjoy my home.

I'll tell you another thing. Everything I've talked about I can back up. Each and every bit of it. I have saved everything. When we moved to our new house, my husband said, "Ya know, moving all that stuff costs money!"

I said, "Well, we're movin' it."

I wanted to be in drag racing as long as I remained competitive. In 2003, I set my quickest and fastest career run at the Route 66 Raceway in Chicago (top). My e.t. was 4.57, and I was clocked at 327 miles an hour. Later that year at Pomona (bottom) I made my last pass in a dragster. *Shirley Muldowney Collection/Dave DeAngelis*

Those are my files with all my records, all my fan mail—you wouldn't believe what I've held on to.

I still get to go to the races, and I couldn't be treated more nicely by everyone. All the team members, the drivers, and the fans treat me like royalty. Of course, Rahn works for Connie Kalitta, and I get a kick out of going out and bringing back 35 lunches for them. (Despite Connie's and my personal squabbles. Rahn and Connie have always been friends. Connie and I mended fences years ago.) If I pass a Starbucks, I'll bring back a big coffee for David Grubnic in the morning. I want to be there to support Rahn, and I know he likes it when I go up to the starting line with him.

I still enjoy talking to the fans, who have always been there for me. There's a special kind of respect, I think, that they extend to me and I can't explain it. But I do appreciate it.

There's a chance I could become a team manager in the near future. The opportunity has presented itself, although nothing is definite at this point. I won't drive a race car again, but I think I know what you need to do to manage a team. I did it for myself for many years. If it happens, it happens. If it doesn't, I'll just keep doing what I've always been able to do.

I'll survive.

Rahn and I are enjoying life now that I have hung up my helmet. *Shirley Muldowney Collection*

Epilogue

Following her retirement from the rigors of NHRA competition in 2003, Shirley Muldowney briefly served as Team Sponsor Relations Manager for the Top Fuel operation fielded by Team Basic Research, the makers of Zantrex3 and StriVectin.

Shirley frequently accepts invitations to make personal appearances at selected racing events, hot rod shows, NHRA functions, and enthusiast conventions to sign autographs and meet with the huge legion of fans who still recall and appreciate her contributions to the sport of drag racing.

She and her second husband, Rahn Tobler, divorced in 2006.

In 2007, her first husband, Jack Muldowney, passed away at which time Shirley remarked, "I owe so much to Jack and I have always been very grateful for everything he did that led to my success. He was a wonderful man."

Since her 2003 retirement, Shirley has received numerous offers for her to return to the driver's seat, either in a Top Fuel Dragster, Nostalgia Funny Car, or most recently, a 500-plus MPH land speed record car to set a new land speed record for women. "The right deal hasn't come along just yet," says Muldowney. "But if one should materialize, I'm still ready and willing to race again, and hopefully, serve as a mentor to other female racers looking to pursue a professional racing career."

Shirley resides in her suburban Michigan home with her beloved dogs while enjoying gardening, cooking, and answering the many letters and emails she still receives from her devoted fans.

On June 19, 2013, Shirley celebrated her 73rd birthday.

APPENDIX

Career Statistics

NHRA Top Fuel Champion 1977, 1980, 1982

AHRA Top Fuel Champion 1981

First and only woman to win a professional racing championship

First Top Fuel driver to win two NHRA championships

No. 5 on NHRA's list of 50 Greatest Drivers in History

Won 18 NHRA national events out of 27 final rounds

Scored 13 NHRA No. 1 qualifying spots

Scored Low Elapsed Time at 12 NHRA national events

Scored Top Speed at 19 NHRA national events

Beat Connie Kalitta more than any other driver in her career, 15 times

Inductee into four Halls of Fames

NHRA National Event Wins

1976 Springnationals, Columbus, Ohio
Beat Bob Edwards in the final round

1976 World Finals, Ontario, California
Beat Jerry Ruth in the final round

1977 Springnationals, Columbus, Ohio
Beat Clive Skelton in the final round

1977 Summernationals, Englishtown, New Jersey
Beat Jeb Allen in the final round

1977 LeGrand Nationals, Montreal, Quebec
Beat Pat Dakin in the final round

1980 Winternationals, Pomona, California
Beat Connie Kalitta in the final round

1980 Springnationals, Columbus, Ohio
Beat Larry Brown in the final round

1980 NHRA Nationals, Seattle, Washington
Beat Marvin Graham in the final round

1980 World Finals, Ontario, California
Beat Frank Bradley in the final round

1981 Gatornationals, Gainesville, Florida
Beat Jody Smart in the final round

1981 Southern Nationals, Atlanta, Georgia
Beat Terry Capp in the final round

1981 NHRA Nationals, Irvine, California
Beat Joe Amato in the final round

1982 Gatornationals, Gainesville, Florida
Beat Don Garlits in the final round

1982 Springnationals, Columbus, Ohio
Beat Lucille Lee in the final round

1982 NHRA Nationals, Brainerd, Minnesota
Beat Gary Beck in the final round

1982 U.S. Nationals, Indianapolis, Indiana
Beat Connie Kalitta in the final round

1983 Winternationals, Pomona, California
Beat Jody Smart in the final round

1989 NHRA Nationals, Phoenix, Arizona
Beat Darrell Gwynn in the final round

Additional Career Milestones

- During 1990s, set 12 new track records at various match-racing events in North America
- In 1993, set new track record at Fuji International Speedway in Japan: 5.30-second pass at 285 mph
- In 1996, qualified in the top three at every IHRA national event; advanced to five consecutive final rounds, winning three and finishing second in the final Top Fuel standings
- In 1997 set new IHRA Top Speed national record four times
- In 1998 set all-time quickest and fastest IHRA pass at the time at Northern Nationals in Stanton, Michigan: 4.69-second pass at 312.50 mph; set track records at three other events
- In 2000 won Millennium Race against "Big Daddy" Don Garlits at Moroso Motorsports Park in West Palm Beach, Florida
- In 2001 at Mac Tools U.S. Nationals, ran career-best numbers at that time: 4.64-second pass at 320.20 mph
- In 2003 during the Last Pass tour at NHRA Nationals in Joliet, Illinois, ran all-time career-best elapsed time and Top Speed: 4.578-second pass at 327.66 mph

Statistics provided by Bob Frey